COMBUSTION-GENERATED
AIR POLLUTION

COMBUSTION-GENERATED AIR POLLUTION

A Short Course on Combustion-Generated Air Pollution
held at the University of California, Berkeley
September 22-26, 1969

Edited by Ernest S. Starkman

Department of Mechanical Engineering
University of California
Berkeley, California

℗ PLENUM PRESS · NEW YORK – LONDON · 1971

Library of Congress Catalog Card Number 73-155925
ISBN-13: 978-1-4684-7576-0   e-ISBN-13: 978-1-4684-7574-6
DOI: 10.1007/978-1-4684-7574-6
© 1971 Plenum Press, New York
Softcover reprint of the hardcover 1st edition 1971
A Division of Plenum Publishing Corporation
227 West 17th Street, New York, N.Y. 10011

United Kingdom edition published by Plenum Press, London
A Division of Plenum Publishing Company, Ltd.
Davis House (4th Floor), 8 Scrubs Lane, Harlesden, NW10, 6SE, England

CONTENTS

CONTRIBUTORS

Laurence S. Caretto, Department of Mechanical Engineering,
    University of California at Berkeley, Berkeley, California

Ellis F. Darley, Statewide Air Pollution Research Center,
    University of California, Riverside, California

James G. Edinger, Professor of Meteorology, University of
    California, Los Angeles, California

Milton Feldstein, Bay Area Air Pollution Control District, San
    Francisco, California

John A. Maga, California Air Resources Board, Sacramento,
    California

Peter K. Mueller, Chief, Air and Industrial Hygiene Laboratory,
    State Department of Public Health, Berkeley, California

Jay A. Nadel, Cardiovascular Research Institute, University of
    California School of Medicine, San Francisco, California

Victor P. Osterli, Agricultural Extension Service, University of
    California at Davis, Davis, California

James N. Pitts, Jr., Professor of Chemistry, University of
    California, Riverside, California

Robert F. Sawyer, Associate Professor of Mechanical Engineering,
    University of California, Berkeley, California

Ernest S. Starkman, University of California, Berkeley, California

INTRODUCTION

    This collection of notes was assembled as a supplement and guide
to a five-day short course presented at the University of California
at Berkeley, September 22-26, 1969. The scope of subject matter,
while limited to combustion as a source of air pollution, at the
same time is intended to give the broadest possible exposure within
that area. The spectrum is deliberately wide, ranging from fundamentals of combustion and combustion reactions through performance of
combustion systems and to legal and administrative control.

    Contributors to this compendium and lecturers in the subject were
solicited from academic and public organizations. Most of the authors
are from the statewide University of California and the California
Department of Public Health. Notable individuals with particular ex-
pertise, from other institutions, were also invited to contribute.
The choice of instructor in each case was based upon a desire to collect a cross-section of outstanding individuals, each highly qualified
technically in his field. These notes reflect the freedom which each
author was encouraged to follow in providing supplementary material
for his lecture.

    The staff of Continuing Education in Engineering, Professor Thomas
Hazlett and Daphne Stern, deserve commendation for their effective and
successful handling of the innumerable details which were encountered.
Professors Robert Sawyer and Laurence Caretto are herewith gratefully
acknowledged for their support in the seemingly uncountable tasks necessary to assemble the entity which is represented.

    The material here outlined is intended to be only a beginning. It
is the desire to expose each individual to the fundamentals and practi-
calities of combustion-generated air pollution as it is understood today. The objective is to provide to each thereby a better array of
equipment to be used in understanding and attacking the problems of
air pollution.

Ernest S. Starkman

Berkeley, California

COMBUSTION THERMODYNAMICS

Laurence S. Caretto and Robert F. Sawyer
Department of Mechanical Engineering
University of California at Berkeley
Berkeley, California

INTRODUCTION

The majority of man's energy long has become available through combustion processes. In the United States over 95% of the energy consumed involves the combustion of fossil fuels (1). On a world-wide basis, more than 99% of the energy consumed is released through combustion processes (2). While increased development of nuclear power will occur during the remainder of this century, fossil fuels are predicted to provide more than 75% of the energy requirements of the United States in the year 2000. In 35 years the consumption of fossil fuels will be 2 to 3 times the current rate (1).

In a general sense, combustion is taken to include all fast exothermic reations. "Fast" implies that the time scale is of the order of seconds, or less. More often, "fast" means milliseconds or less. "Exothermic," in this same vague sense, implies that the energy involved is sufficient to be of interest.

Recent intensification of the study of combustion has been associated with advent and development of aerospace propulsion, primarily rocketry. While the problems of space and atmosphere propulsion will continue to place new demands upon the field of combustion, the primary requirement for improved understanding of combustion is likely to result from air pollution problems.

Many of the problems of understanding and describing the nature of combustion remain unsolved. This lack of progress results from the intrinsic complexity of combustion phenomena rather than from a lack of interest or effort. These complexities are traceable directly to the result of the simultaneous interaction in the combustion process of (1) chemical reaction, (2) fluid flow, (3) heat transfer, and (4) mass transfer.

It is natural that the work in combustion has been carried on by a group in which chemists, thermodynamicists, and aerodynamicists are represented. The resulting mix of metric and English dimension persists in the combustion literature. No attempt is made here to promote the consistent use of a particular system.

LAWS OF THERMODYNAMICS

Pertinent laws of thermodynamics and basic thermodynamic definitions and concepts are presented in their simplest forms. The material is recorded in the form of a review. If a more detailed or rigorous presentation is sought, a number of suitable references exist, for example (3) or (4).

Equation of state: Combustion products generally are high temperature gases and as such may be treated adequately as perfect gases, i.e. fluids obeying the law, $PV = nRT$. A corollary relation for a perfect gas is that its internal energy be a function of temperature alone, $U = U(T)$ alone. The term "calorically perfect gas" is reserved for a gas with a constant specific heat, $C_V = $ constant.

Work: Work in a combustion system is limited to mechanical work of the form, $đW = P\,dV$. The amount of work produced by a system depends upon the path taken, in terms of, say, the thermodynamic coordinates P and V. Work, therefore, is not a function of the thermodynamic coordinates of the initial and final states of the system alone. The differential quantity of work is an inexact differential and represented by the notation, $đW$, rather than dW which is reserved for an exact differential.

Heat: The term heat refers to energy in transit because of a temperature difference. Since heat cannot be expressed as a function of the thermodynamic coordinates, the infintesimal quantity of heat is an inexact differential and therefore represented as $đQ$.

First law of thermodynamics: An expression of conservation of energy, the first law of thermodynamics merely states that the difference between the heat added to a system and the work produced by the system is the change in the associated internal energy function, U.

$$đQ = dU + đW \tag{1}$$

The sign convention adopted is that heat transferred to the system and work done by the system are positive. In a combustion system, work is limited to PdV work so that the first law of thermodynamics takes the form $đQ = dU + PdV$. The equivalence between mechanical energy and thermal energy is expressed as 4.1858 joules/cal or 778 ft-lb/Btu.

Heat capacity: The heat capacity of a system is defined by the ratio of heat transfer to resultant temperature change, $C = dQ/dT$. The heat capacity takes on unique values for processes occurring at constant pressure and at constant volume. The heat capacity at constant pressure, $C_P = (dQ/dT)_P$, and the heat capacity at constant volume, $C_V = (dQ/dT)_V$. For a perfect gas the difference between these specific heats is a constant, $C_P - C_V = R$. Note, however, that only for calorically perfect gases are the specific heats themselves constant. High temperature gases characteristic of combustion products have specific heats which depend strongly upon temperature and therefore do not qualify as calorically perfect gases. Heat capacities may be measured experimentally at moderate temperatures and calculated with good accuracy at high temperatures from the spectroscopic characteristics of the gas. An extensive tabulation of heat capacities as a function of temperature is available in the JANAF Thermochemical Tables (5).

Second law of thermodynamics: The second law of thermodynamics places limitation upon the amount of work derivable in a thermodynamic process. More important among the consequences of the second law is the relation of heat transfer to the thermodynamic coordinates of the system. For a reversible process,

$$\frac{dQ_{rev}}{T} = dS \qquad (2)$$

The differential change in the entropy, dS, is exact. Equation (2) is a definition for the thermodynamic property, the entropy. For an ideal gas, the entropy change is related in a simple fashion to the other thermodynamic coordinates.

$$dS = C_P \frac{dT}{T} + nR \frac{dP}{P} \qquad (3)$$

$$dS = C_V \frac{dT}{T} + nR \frac{dV}{V} \qquad (4)$$

Reversible processes are those for which the entropy of the system and its surroundings remain unchanged. Reversible adiabatic processes are necessarily isentropic. Chemical reaction and therefore combustion processes are irreversible and accompanied by an entropy increase. In a microscopic consideration of thermodynamics, the entropy is given an additional significance -- entropy is a measure of the disorder of the system.

The most important idea of the second law is that the entropy of an isolated system is a maximum at equilibrium. This principle

allows the prediction of the final equilibrium state that will obtain in a system. If the entropy maximum principle is applied to
a simple system with two parts one finds the expected result that

$$T^{(2)} = T^{(1)}$$

for equilibrium when energy transfer is allowed and

$$P^{(2)} = P^{(1)}$$

for equilibrium when the two subsystems can exchange volume. The
entropy maximum principle can be reformulated in terms of other
thermodynamic functions but the second law always gives an extremum
principle which can be used to find a final equilibrium state.

It is necessary to use such a principle to determine the
equilibrium composition of a reactive mixture.

Enthalpy: The enthalpy of a system is defined by $H = U + PV$. The
entalpy is a particularly useful quantity in the description of
constant pressure processes. The differential of entalpy is exact.
The entalpy, therefore, may be evaluated as the integral of an
analytic function. In particular, for an ideal gas,

$$dH = C_P dT \qquad (5)$$

which may be integrated to give

$$H_T = H_0 + \int_{T_0}^{T} C_P \, dT \qquad (6)$$

As was mentioned previously, the specific heat for high temperature
gases is a function of the temperature. The integral, therefore,
may be evaluated only if the dependence of the specific heat upon
temperature, $C_P = C_p(T)$, is known. Only for the special case of
the calorically perfect gas is $H_T - H_0 = C_P(T - T_0)$.

In chemically reactive systems, the enthalpy arising from
temperature change, the "sensible enthalpy" or $\int C_P \, dT$, is not sufficient to describe completely the enthalpy of the system. To
account for the capacity of the system to undergo chemical reactions
one must identify the constant of integration in equation (6) as
the enthalpy of formation of the specie. In standard notation,
then,

$$H_0 = \Delta H^0_{f_{T_0}} \qquad (7)$$

The enthalpy of formation, $\Delta H^0_{f_{T_0}}$, is the enthalpy required to form the specie from the standard state elements at a pressure of one atmosphere and at a temperature, T_0. By convention, the standard state elements are taken to be the elements in their natural forms. For example, standard state elements are: gaseous molecular nitrogen, $N_2(g)$; gaseous molecular hydrogen, $H_2(g)$; graphitic carbon, $C(gr)$; etc. The reference temperature, T_0 is taken to be 298.15^0K (25^0C. or 77^0F.). The superscript (0) denotes the standard state condition of 1 ATM. The enthalpies of formation for ideal gases are independent of pressure, i.e., functions of temperature alone. Some enthalpies of formation of species of interest to combustion processes are recorded in Table 1.

Enthalpies of formation are measured experimentally by calorimetric means, deduced from measured enthalpies of formation of related species, or estimated from bond energies. The enthalpy of formation is an index of the potential of the specie to undergo an exothermic reaction. The larger the enthalpy of formation, the more likely is the specie to enter into an exothermic reaction. Nearly non-reactive species, such as water or carbon dioxide, have negative enthalpies of formation. Once again, the JANAF Thermochemical Tables (5) provide an excellent source for enthalpies of formation.

Internal energy: The internal energy, U, was first mentioned in connection with the statement of the first law of thermodynamics. For combustion studies, it is generally necessary only to include sensible and chemical energies in the accounting of internal energies. The differential of internal energy is exact. For an ideal gas,

$$dU = C_V dT \tag{8}$$

and may be integrated to give,

$$U_T = U_0 + \int_{T_0}^{T} C_V \, dT \tag{9}$$

As for the enthalpy, the constant of integration is related to a capacity to undergo an exothermic reaction.

$$U_0 = \Delta U^0_{f_{T_0}} \tag{10}$$

other forms of internal energy, e.g., gravitational potential energy and kinetic energy, may be neglected in most combustion

Table 1.   Enthalpies of formation of species of interest as combustion reactants or products. Formation from assigned reference elements, 298.15°K, in kcal/mole. Gas phase unless otherwise noted.

C	171.3	$C_6H_6(l)$	19.8
C(gr)	0	$n-C_6H_{14}$	−40.0
CH	142.4	$n-C_7H_{16}$	−44.9
CHO	−3.2	$n-C_8H_{18}$	−49.8
CH_2	69.0	H	52.1
CH_3	33.4	HN	78.9
CH_3OH	−48.1	HNO	23.8
$CH_3NO_2(l)$	−26.7	$HNO_3(l)$	−41.4
CH_4	−17.9	HO	9.3
$CH_4(l)$	−20.3	HO_2	5.0
CN	96.4	H_2	0
CO	−26.4	H_2O	−57.8
CO_2	−94.1	H_3N	−11.0
C_2	200.0	$H_4N_2(l)$	12.05
C_2H_2	54.9	N	113.0
C_2H_4	12.5	NO	21.6
C_2H_6	−20.2	NO_2	8.0
C_2H_5OH	−50.2	N_2	0
C_2N_2	73.8	N_2O	19.5
C_3	189.7	N_2O_4	2.3
C_3H_8	−24.8	O	59.6
$C_3H_7OH(l)$	−71.9	O_2	0
$n-C_5H_{12}$	−35.0		

problems. Should the need arise, then any other form of internal
energy could be included in the "energy accounting system" described above.

Helmholtz and Gibbs functions: These two functions, often called
the Helmholtz free energy and the Gibbs free energy, are used to
determine equilibrium conditions for systems under specified constraints. The Helmholtz function, A, is defined as

$$A = U - TS \tag{11}$$

and obeys the following extremum principle: the Helmholtz function
is a minimum for the final equilibrium state of a system at constant volume and constant temperature.

The Gibbs function, G, is defined as

$$G = H - TS \tag{12}$$

the Gibbs function is a minimum for the final equilibrium state of
a system at constant temperature and pressure.

The symbol F is commonly used for the Helmholtz function by
physicists and for the Gibbs function by chemists. (The JANAF
tables use F for the Gibbs function).

The (exact) differentials of the various thermodynamic
properties are given by the following relations:

$$dU = TdS - PdV \tag{13}$$

$$dH = TdS + VdP \tag{14}$$

$$dA = -PdV - SdT \tag{15}$$

$$dG = VdP - SdT \tag{16}$$

Mixtures: It is convenient to relate the thermodynamic properties
of a mixture to the thermodynamic properties of its components.
To do this in the general case requires the use of so-called partial

molar quantities (4). Fortunately most combustion problems involve
high temperature gases which are adequately described by ideal gas
laws. For ideal gases the mixture property is simply the weighed
sum of the component properties at the same temperature and partial
pressure of the component in the mixture. The partial pressure,
p_i, is defined as

$$p_i = x_i P = (n_i/n)P \tag{17}$$

where x_i is the mole fraction of species i, n_i is the number of
moles of species i, and n is the total number of moles; obviously,

$$\sum_{i=1}^{m} n_i = n \tag{18}$$

$$\sum_{i=1}^{m} x_i = 1 \tag{19}$$

$$\sum_{i=1}^{m} p_i = P \tag{20}$$

where m is the number of species in the mixture.

The internal energy of the ideal gas mixture is given by

$$U(T,x) = \sum_{i=1}^{m} n_i u_i(T) \tag{21}$$

and the mixture entropy is given by

$$S(T,P,x) = \sum_{i=1}^{m} n_i x_i(T,p_i) \tag{22}$$

The lower case symbol denotes the quantity per mole. Note that
the entropy of a species must be evaluated at the partial pressure
of that species. The temperature and total pressure are uniform
throughout an equilibrium thermodynamic system.

Composition change. For open systems and systems where the composition changes due to chemical reactions equations 13-16 must be
augmented to account for the effects of composition change on the
thermodynamic properties. This is done by introducing the chemical
potential defined as

$$\mu_i = (\frac{\partial U}{\partial n_i})_{S,V,n_{j\neq i}} = (\frac{\partial H}{\partial n_i})_{S,P,n_{j\neq i}} = (\frac{\partial A}{\partial n_i})_{T,V,n_{j\neq 1}} = (\frac{\partial G}{\partial n_i})_{T,P,n_{j\neq 1}}$$

The notation $n_{j\neq i}$ implies that the differentiation with respect to n_i is taken holding all the other moles numbers n_j ($j\neq i$) constant. With this definition the differentials of the thermodynamic properties become

$$dU = TdS - PdV + \sum_{i=1}^{m} \mu_i dn_i \qquad (23)$$

$$dH = TdS + VdP + \sum_{i=1}^{m} \mu_i dn_i \qquad (24)$$

$$dA = PdV - SdT + \sum_{i=1}^{m} \mu_i dn_i \qquad (25)$$

$$dG = VdP - SdT + \sum_{i=1}^{m} \mu_i dn_i \qquad (26)$$

Note that these equations are completely general and reduce to equations 13-17 for constant composition (all $dn_i = 0$).

The chemical potential is a somewhat abstract quantity which is best understood as the mass transfer analog of the temperature. We consider heat to flow from higher to lower temperatures and thermal equilibrium to be reached when temperatures are equal. Similarly, if it is possible for matter of species i to be transferred between two systems, the matter will flow from the system with the higher value of chemical potential, μ_i, to that with the lower value of μ_i, and equilibrium with respect to mass transfer of species i occurs when the value of μ_i is the same for both systems.

STOICHIOMETRY

When chemical reactions take place all the mole number changes are no longer independent. The conservation of chemical elements places restrictions on the allowed variations of the dn_i. Analysis of the composition change in a chemical reactions comes under the name of "stoichiometry." It is important to distinguish between the actual composition change in a reactive process and the use of balanced chemical reactions which are a tool used in solving problems.

As an example consider the reaction of hydrogen and oxygen. The possible species include H_2, O_2, H_2O, OH, H, and O. (It is also possible to consider peroxy species HO_2 and H_2O_2, but these will be neglected in this example.) We have 6 species to consider in this case, but not all are independent since we must have conservation of H atoms and O atoms, i.e.

$$n_H + 2n_{H_2} + 2n_{H_2O} + n_{OH} = \text{constant}$$

$$n_{H_2O} + n_{OH} + n_O + 2n_{O_2} = \text{constant}$$

The two equations of constraint reduce the number of independent composition variables from six to four. We could differentiate the two equations immediately above and obtain a relation between the dn_i which would allow us to eliminate two of them. The general procedure, which is completely equivalent, is to consider balanced chemical equations. One then defines a reaction progress variable for each equation. In this case we need four independent variables so we would require four reactions, e.g.

$$1) \qquad H_2 = 2H$$

$$2) \qquad O_2 = 2O$$

$$3) \quad H_2 + O_2 = 2OH$$

$$4) \quad 2H_2 + O_2 = 2H_2O$$

The composition changes are then related to the degrees of advancement, ξ, of the chemical reactions as follows

$$dn_H = 2d\xi_1$$

$$dn_O = 2d\xi_2$$

$$dn_{OH} = 2d\xi_3$$

$$hd_{H_2O} = 2d\xi_4$$

$$dn_{H_2} = -d\xi_1 - d\xi_3 - 2d\xi_4$$

$$dn_{O_2} = -d\xi_2 - d\xi_3 - d\xi_4$$

The above equations indicate that we are effectively regarding dn_H, dn_O, dn_{OH}, and dn_{H_2O} as the independent variables in this system. The use of balanced chemical reactions provides a straightforward, general scheme for taking atom conservation into account. (This is to be expected as atom conservation forms the basis for balanced chemical reactions.)

In the general case we can write our reactions as

$$\sum_{i=1}^{m} \nu'_{ij} R_i = \sum_{i=1}^{m} \nu''_{ij} P_i \qquad j=1\ldots r \qquad (27)$$

where R_i denotes a reactant species, P_i denotes a product species and the quantities ν'_{ij} and ν''_{ij} are called the stoichiometric coefficients for reaction j. The number of reactions, r, which must be considered is simply equal to the number of chemical species less the number of different atoms present in the system. Any choice of linearly independent chemical reactions will suffice for a thermodynamic analysis. It is usually easiest to choose the set of chemical reactions where species are formed from their naturally occurring chemical elements. This provides a consistent scheme which is readily adapted to machine computation.

In terms of the reaction progress variables for our general scheme the changes in mole number are given by

$$dn_i = \sum_{j=1}^{r} (\nu''_{ij} - \nu'_{ij})\, d\xi_i \qquad (28)$$

Note that we formally consider each species to be written on both sides of every chemical equation. This is done to provide easy extension to a computer solution. Of course, most of the stoichiometric coefficients will be zero.

In the example of hydrogen-oxygen reaction we have the following tables:

ν'_{ij}

species i \ reaction j	1	2	3	4
H	0	0	0	0
O	0	0	0	0
OH	0	0	0	0
H_2O	0	0	0	0
H_2	1	0	1	2
O_2	0	1	1	1

ν''_{ij}

species i \ reaction j	1	2	3	4
H	2	0	0	0
O	0	2	0	0
OH	0	0	2	0
H_2O	0	0	0	2
H_2	0	0	0	0
O_2	0	0	0	0

From equation 28 we can write

$$\sum_{i=1}^{m} \mu_i dn_i = \sum_{i=1}^{m} \mu_i \sum_{j=1}^{r} (\nu''_{ij} - \nu'_{ij}) d\xi_j =$$

$$\qquad\qquad (29)$$

$$\sum_{i=1}^{r} (\sum_{i=1}^{m} (\nu''_{ij} - \nu'_{ij}) \mu_i) d\xi_j$$

This can be substituted into equations 23-26. In particular we can write dG as

$$dG = SdT + VdP + \sum_{j=1}^{r} (\sum_{i=1}^{m} (\nu''_{ij} - \nu'_{ij}) \mu_i) d\xi_j \qquad (30)$$

The fundamental difference between the mole numbers, n_i, which represent the actual composition in a given problem, and the stoichiometric coefficients, ν'_{ij} and ν''_{ij}, which represent the mathematical coefficients in a balanced chemical equation should be clearly recognized. The stoichiometric coefficients are simply numerical tools which allow us to calculate the composition.

Specification of the initial reactant composition usually is made in terms of mixture ratio. A quantity such as the fuel-oxidizer mass ratio may be used. The equivalence ratio is a particularly useful quantity to specify the reactant composition. Defined as

$$\phi = \frac{\dfrac{fuel}{oxidizer}}{\left(\dfrac{fuel}{oxidizer}\right)_{stoichiometric}} \qquad (31)$$

the equivalence ratio is a normalized, non-dimensional quantity and may therefore be formed from the mass, mole, or volume ratios of the fuel and oxidizer. The stoichiometric mixture ratio corresponds to complete reaction, that is, to the reactant quantities which would allow reaction to normal oxidation state products. (For systems with C, H, O, and N atoms only the stoichiometric reactant composition is the one which could produce only CO_2, H_2O, and N_2 on the basis of atom conservation.) An equivalence ratio of less than one is fuel lean; greater than one, fuel rich; and equal to one, stoichiometric. For the hydrogen/oxygen reaction

$$\phi = (n_{oH_2}/n_{oO_2})/(1/\tfrac{1}{2}) = (m_{oH_2}/m_{o\,O_2})/(4/32)$$

(m = mass; subscript ($_o$) denotes initial composition).

THERMOCHEMISTRY

In calculations of changes of the thermodynamic properties in reacting systems one can simply write

$$Y_2 - Y_1 = \sum_{i=1}^{m} n_{i2} y_i (T_2 \ P_2 \ n_2) - \sum_{i=1}^{m} n_{i1} y_i (T_1, P_1, n_1) \quad (32)$$

where Y denotes any extensive thermodynamic property. In particular the enthalpy change for a process is given by

$$H_2 - H_1 = \sum_{i=1}^{m} n_{i_2} h_i (T_2) - \sum_{i=1}^{m} n_{i_1} h_i (T_1) \quad (33)$$

where the enthalpy is given by equations 6 and 7.

One also speaks of the standard _enthalpy of reaction_ which is defined for a balanced chemical reaction, at a given temperature; this is defined as

$$\Delta H^\circ_{R_j} (T) = \sum_{i=1}^{m} (\nu''_{ij} - \nu'_{ij}) \ h^\circ_i (T) \quad (34)$$

again the enthalpies are given by equations 6 and 7. Note the difference between the two definitions of an enthalpy change; one involves the actual change in mole numbers and two thermodynamic states; the other involves the stiochiometric coefficients and has both reactants and products at the same temperature and standard state pressure.

The values of ΔH_R for balanced stoichiometric combustion reactions are usually given for the reaction of 1 mole of fuel (i.e. $\nu'_{fuel} = 1$). The general stoichiometric reaction for a fuel with some combination of C, H. O, or N. atoms is

$$1 \ \text{Fuel} + \nu'_{O_2} O_2 = \nu''_{CO_2} CO_2 + \nu''_{H_2O} H_2O + \nu''_{N_2} N_2$$

for which (subscript F denotes fuel)

$$\Delta H^\circ_R (T_o) = \nu_{CO_2}'' h_{CO_2}(T_o) + \nu_{H_2O}'' h_{H_2O}(T_o) + \nu_{N_2}'' h_{N_2}(T_o)$$

$$- \nu'_{O_2} h_{O_2}(T_o) - h_F(T_0) \quad (35)$$

Combustion reactions are by definition always exothermic. The enthalpy of formation of the reactants must therefore exceed the enthalpy of formation of the products. From equation (35), en-

thalpies of reaction for combustion reactions necessarily have
negative signs. High enthalpies of reaction are associated with
low enthalpy of formation product species and high enthalpy of
formation reactants. Enthalpies of reaction the negative of
which is sometimes referred to as "heat of combustion") for some
combustion reactions of interest are presented in Table 2.

If the final temperature in a combustion process is low
enough so that dissociation effects are not important the actual
enthalpy change can be written in terms of $\Delta H^\circ_R(T_0)$. If we have
$n_{O_{2i}}$ moles of O_2 initially we will have $b\ n_{O_{2i}}$ moles of N_2
initially where

$$b = \begin{cases} 0 \text{ for pure oxygen} \\ 3.76 \text{ for air} \end{cases}$$

We can write the overall process as

$$[n_F \text{ Fuel} + n_{O_{2i}} (O_2 + b\ N_2)] \text{ at } T=T_i$$

$$\rightarrow [n_{CO_2} CO_2 + n_{N_2} N_2 + n_{H_2O} H_2O + n_{O_2\ f} O_2] \text{ at } T=T_f$$

from equation 33 we can write

$$\frac{H_f - H_i}{n_F} = \frac{n_{CO_2}}{n_F} h_{CO_2}(T_f) + \frac{n_{H_2O}}{n_F} h_{H_2O}(T_f) + \frac{n_{N_2}}{n_F} h_{N_2}(T_f)$$

$$+ \frac{n_{O_2}}{n_F} h_{O_2}(T_f) - \frac{n_{O_{2i}}}{n_i} h_{O_2}(T_i)$$

$$- \frac{b n_{O_{2i}}}{n_F} h_{N_2}(T_i) - h_F(T_i)$$

From the definition of equivalence ratio (equation 31) we can
write

$$\frac{n_{O_{2i}}}{n_F} = \frac{\nu'_{O_2}}{\Phi}$$

and since ν'_{O_2} moles of O_2 must react for each mole of fuel we
have

$$\frac{n_{O_{2f}}}{n_F} = \nu'_{O_2}\left(\frac{1}{\Phi} - 1\right)$$

Since $\Phi \leq 1$ for complete combustion, this term is always positive.
Also by stiochiometry we have

Table 2. Enthalpies of reaction for some typical combustion reactions. Oxidation by oxygen to water and carbon dioxide at one atmosphere pressure

Fuel	Phase	Temperature [°C]	$-\Delta H_R$ [kcal/mole fuel]*
C	graphite	25	94.0
CH_4	gas	20	210.8
CH_3OH	liquid	20	170.9
CO	gas	25	67.7
C_2H_2	gas	20	312.0
C_2H_4	gas	20	331.6
C_2H_6	gas	20	368.4
C_2H_5OH	liquid	20	327.6
$n-C_5H_{12}$	liquid	20	833.4
C_6H_6	liquid	20	782.3
$n-C_6H_{14}$	liquid	20	989.8
$n-C_7H_{14}$	liquid	20	1149.9
$n-C_8 H_{18}$	liquid	20	1302.7
H_2	gas	25	57.8
NH_3	gas	25	75.7
N_2H_4	liquid	25	127.6

*It is possible to define two values of $-\Delta H_R$ because of the two possible phases of the product H_2O. These figures are based on liquid H_2O as the product. $(-\Delta H_R)$ for H_2O vapor as product = $(-\Delta H_R)$ for H_2O liquid as product $- \nu''_{H_2O} \lambda_{H_2O}(T)$, where λ_{H_2O} = latent heat = 1.055 kcal/mole at 20°C, 1.050 kcal/mole at 25°C.

$$\frac{n_{CO_2}}{n_F} = \nu''_{CO_2}$$

$$\frac{n_{H_2O}}{n_F} = \nu''_{H_2O}$$

$$\frac{n_{H_2}}{n_F} = \nu''_{N_2} + \frac{bn_{O_2i}}{n_F} = \nu''_{N_2} + \frac{b\nu'_{O_2}}{\Phi}$$

Substituting these expressions for moles/mole of fuel into the equation for enthalpy change and writing

$$h[T] = h[T_o] + \int_{T_o}^{T} c_p dT$$

gives

$$\frac{H_f - H_i}{n_F} = \nu''_{CO_2}\ [h_{CO_2}[T_o] + \int_{T_o}^{T_f} c_{P_{CO_2}}\ dT]$$

$$+ \nu''_{H_2O}\ [h_{H_2O}(T_o) + \int_{T_o}^{T_f} c_{P_{H_2O}}\ dT]$$

$$+ \nu'_{O_2}\ (\frac{1}{\Phi} - 1)\ [h_{O_2}\ (T_o) + \int_{T_o}^{T_f} c_{P_{O_2}}\ dT]$$

$$+ (\nu''_{N_2} + \frac{b\nu'_{O_2}}{\Phi})\ [h_{N_2}(T_o) + \int_{T_o}^{T_f} c_{P_{N_2}}\ dT]$$

$$- \frac{\nu'_{O_2}}{\Phi}\ [h_{O_2}\ (T_o) + \int_{T_o}^{T_f} c_{P_{O_2}}\ dT] - [h_F(T_o) + \int_{T_o}^{T_i} c_{P_F}\ dT]$$

$$- \frac{b\nu'_{O_2}}{\Phi}\ [h_{N_2}\ (T_o) + \int_{T_o}^{T_i} c_{P_{N_2}}\ dT]$$

which can be rearranged to give

$$\frac{H_f - H_i}{n_f} = \Delta H_R^o(T_o)$$

$$+ \int_{T_o}^{T_f} [\nu_{CO_2}'' \; c_{P_{CO_2}} + \nu_{H_2O}'' \; c_{P_{H_2O}} + \nu_{O_2}' (\frac{1}{\Phi} - 1) c_{P_{O_2}} + \nu_{N_2}'' \; c_{P_{N_2}}] dT$$

$$+ \frac{\nu_{O_2}' b}{\Phi} \int_{T_i}^{T_f} c_{P_{N_2}} dT$$

$$- \int_{T_o}^{T_i} (c_{P_F} + \frac{\nu_{O_2}}{\Phi} c_{P_{O_2}}) dT$$

where the definition of $\Delta H_R^o (T_o)$ for a combustion reaction (equation 35) has been used. In the special case of $T_f = T_i = T_o$ this reduces to $\Delta H_R^o(T_o)$. We see, then, that ΔH_R^o represents the enthalpy change that would occur if the balanced stoichiometric reaction actually occurred as written.

In many cases the actual process is not complete and disassociation products can significantly affect the calculation of $H_2 - H_1$. In this case it is necessary to use equation 33 which is completely general.

REACTANTS AND REACTION PRODUCTS

Most reactants of concern to air pollution consist of hydrocarbon fuel and air oxidizer. The use of other reactants in combustion processes generally is limited to special applications, for example, rocket propulsion. The presence of small quantities of other compounds, for example. sulphur and lead compounds, may have an important effect on the pollution characteristics of the combustion products and the rate of the combustion process. For purposes of the present discussion, however, the overall combustion process is considered neglecting the small concentrations of any such compounds which are likely to be found in the reactants.

The product species predicted from equilibrium considerations form a relatively short list for hydrocarbon/air reactions. Including minor species which often are present in concentrations of less than one part per million, one may list the following combustion products.

H	OH	HO_2	CO_2	N	NH_3
CH	H_2	C	O	NH	NO
CHO	H_2O	CO	O_2	N_2	NO_2

COMPLEX EQUILIBRIA

The composition of the combustion products may be predicted from equilbrium considerations. The condition of equilibrium for a system at fixed pressure and temperature is that the Gibbs free energy be a minimum. A general technique for determination of equilibrium compositions will be outlined below.

We now consider the conditions for chemical equilibrium. In equation 30 for dG

$$dG = -SdT + VdP + \sum_{j=1}^{r} \left(\sum_{i=1}^{m} (\nu''_{ij} - \nu'_{ij}) \mu_i \right) d\xi_j \qquad (30)$$

We can set $dT = dP = 0$ for constant temperature and constant pressure processes. For such processes, however, we must have G a minimum at equilibrium. This requires $dG = 0$ and hence

$$\sum_{j=1}^{r} \left(\sum_{i=1}^{m} (\nu''_{ij} - \nu'_{ij}) \mu_i \right) d\xi_j = 0$$

But we have said that the $d\xi_j$ are independent variables. Hence the above equation is satisfied if and only if

$$\sum_{i=1}^{m} (\nu''_{ij} - \nu'_{ij}) \mu_i = 0 \qquad\qquad j=1\ldots r \qquad (36)$$

This set of r equations along with the m-r atoms balance equations serves to determine the equilibrium composition. Note that the concept of an equilibrium composition is a direct result of the second law.

To use equation 36 we need an expression for μ_i. For ideal gases we have (4)

$$\mu_i = \mu_i^\circ (T) + RT \ln p_i \qquad (p_i \text{ in atm.})$$

where $\mu_i^\circ(T)$ denotes the standard state (P=1 atm.) value (at temperature T). Values of μ_i° are given in various tables. The JANAF tables (5) use the notation F° for μ_i°.)

Substituting this expression for μ_i into equation 36 gives

$$-\sum_{i=1}^{m} (\nu''_{ij} - \nu'_{ij}) \frac{\mu_i^\circ}{RT} = \sum_{i=1}^{m} (\nu''_{ij} - \nu'_{ij}) \ln p_1 - \ln \prod_{i=1}^{m} p_i^{(\nu''_{ij} - \nu'_{ij})}$$

If we define

$$\Delta G^{\circ}_{j} = \sum_{i=1}^{m} (\nu''_{ij} - \nu'_{ij}) \mu_{i}^{\circ} \qquad (37)$$

and

$$K_{P_j} = \prod_{i=1}^{m} P_i^{(\nu''_{ij} - \nu'_{ij})} \qquad (38)$$

we have

$$\Delta G^{\circ}_{j} = -RT \ln K_{P_j} \qquad (39)$$

The K_{P_j} contains composition variables; the ΔG°_{j} contains thermodynamic data. The JANAF tables (5) also tabulate the K_P's formation of 1 mole of the species of the species in the table from its elements. For example the K_P column for H_2O contains the equilibrium constant for the reaction

$$H_2 + \frac{1}{2} O_2 = H_2O$$

The tabulated K_P's for O atoms are for the reaction

$$\frac{1}{2} O_2 = O$$

The JANAF K_P's can be used directly if one uses only formulation reactions in formulating a problem. Alternatively it can be shown that the log Kp for any reaction can be written in terms of the JANAF formation K_P's (K_{P_f}) as follows

$$\log K_{P_j} = \sum_{i=1}^{m} (\nu_{ij}'' - \nu_{ij}') \log K_{P_{fi}} \qquad (41)$$

To determine the equilibrium composition at a given temperature and pressure it is necessary to specify r independent chemical reactions and determine their K_P's at the given temperature and tabulation data. From the definitions of Kp and partial pressure we can write

$$K_P = \prod_{i=1}^{m} \left(\frac{n_i}{n} P\right)^{\nu_i'' - \nu_i'}$$

or

$$K_n = \prod_{i=1}^{m} n_i^{\nu_i'' - \nu_i'} = K_P \left(\frac{n}{P}\right)^{\Sigma(\nu_i'' - \nu_i')} \tag{42}$$

The equilibrium composition is specified by the following set of equations

Mass conservation: These equations are equal in number to the number of elements in the products. They express the conservation of elemental species.

Equilibrium: The number of independent equilibrium equations which must be written is equal to the number of species considered less the number of elements contained in these species. These equations take the form of equation 42.

Total number of moles: If the product composition is expressed in terms of moles, then an equation for the total number of moles must be written.

For the example of the hydrogen/oxygen reaction discussed above the equations describing the equilibrium composition are the following:

mass conservation (2 equations)

$$2n_{H_2}' = n_H + n_{OH} + 2n_{H_2} + 2n_{H_2O}$$

$$2n_{O_2}' = n_O + n_{OH} + 2n_{O_2} + n_{H_2O}$$

equilibrium (4 equations)

$$\frac{1}{2} H_2 \rightleftharpoons H \qquad\qquad \frac{n_H}{n_{H_2}^{1/2}} = K_{P_H} \left(\frac{n}{P}\right)^{1/2}$$

$$\frac{1}{2} H_2 + \frac{1}{2} O_2 \rightleftharpoons H \qquad\qquad \frac{n_{OH}}{n_{H_2}^{1/2} n_{O_2}^{1/2}} = K_{P_{OH}}$$

$$H_2 + \frac{1}{2} O_2 \rightleftharpoons H_2O$$

$$\frac{n_{H_2O}}{n_{H_2} n_{O_2}^{1/2}} = K_{P_{H_2O}} \left(\frac{n}{P}\right)^{-1/2}$$

$$\frac{1}{2} O_2 \rightleftharpoons H$$

$$\frac{n_O}{n_{O_2}^{1/2}} = K_{P_O} \left(\frac{n}{P}\right)^{1/2}$$

total number of moles (1 equation)

$$n = n_H + n_{OH} + n_{H_2} + n_{H_2O} + n_O + n_{O_2}$$

The seven preceding equations determine the composition in terms of the seven unknowns, n_H, n_{OH}, n_{H_2O}, n_O, n_{O_2}, n_{H_2}, and n. The solution of these equations in general is difficult. Current practice is to use a computer solution based upon a linear approximation and elimination technique, see, for example, (6, 7). The results of such a machine solution for the composition of a stoichiometric mixture of hydrogen and oxygen at a pressure of one atmosphere and a temperature of 4000°K are presented in Table 3.

The effects of thermodynamic variables on equilibrium compositions follow some general principles typified by some examples here.

Effect of equivalence ratio
=====

The equivalence compositions of hydrocarbon/air mixtures vary greatly depending upon whether the reactants are fuel rich or fuel lean. The reaction products for the combustion of propane and air are presented in Table 4 as a function of equivalence ratio for a fixed temperature. Several general features are noted that relate to the contribution of such combustion products to air pollution. Equilibrium considerations predict no unburned hydrocarbon. The maximum concentration of nitric oxide occurs at the lowest equivalence ratio. The dominant species are the expected stoichiometric products, CO_2, H_2O, and, N_2 plus O_2 at ϕ < 1 and CO and H_2 at ϕ > 1. Other species are present in amounts of less than 1%. While such "small" quantities will have little effect on the energy release, they are important in terms of their potential contributions to air pollution.

Effect of Temperature
=====

Temperature plays a dominant role in fixing the equilibrium composition as is shown in Figure 1. The composition of a

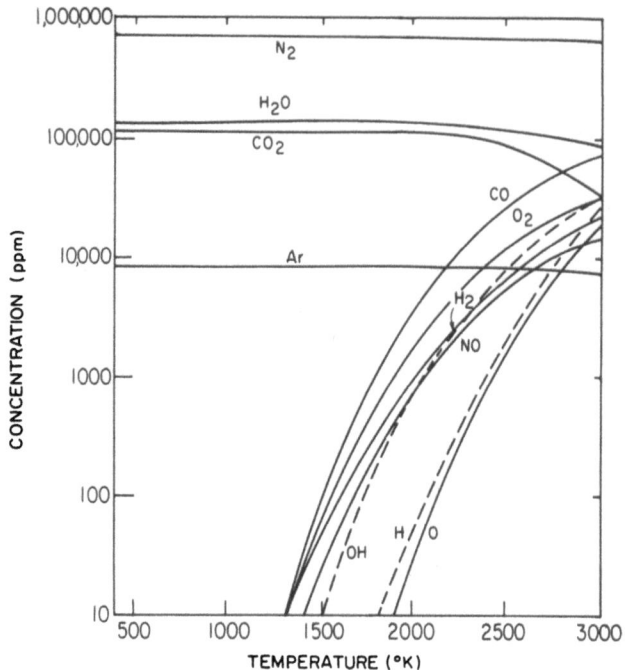

Figure 1. Effect of a temperature on equilibrium composition.
Stoichiometric heptane/air at a pressure of 1 atm.

stoichiometric heptane/air mixture at a pressure of one atmosphere
is presented as a function of temperature. The onset of dissociation of the products of "complete" combustion is seen to occur at
a temperature of about 2000°K.

It can be shown [4] that the temperature variation of the
equilibrium constant is given by

$$\frac{d \ln K_P}{dT} = \frac{\Delta H^{\circ}_R}{RT^2}$$

If $\Delta H_R > 0$ (product enthalpy > reactant enthalpy) K_P, and hence
the product moles, will increase as temperature is increased. We
see then that an increasing temperature tends to favor the formation of higher enthalpy species. Since disociated species have
higher enthalpies than their parent molecules dissociation is
favored by higher temperature.

Effect of pressure

The effect of pressure on equilibrium can arise from two considerations. At a fixed temperature, higher pressures will decrease decompositions and drive the equilibrium composition toward higher molecular weight species, e.g., H_2O and CO_2. As pressure increases, the concentrations of dissociated species, e.g., H, O, and OH, will drop (see Figure 2). For adiabatic combustion the temperature is also affected by the pressure. This second effect is considered in the following section, 2.6.

The effects of pressure (or addition of inert species) at a given temperature, can be deduced from equation 42. For a simple reaction of $A + B \rightarrow C$ this equation becomes $n_C/n_A n_B = (P/n) K_P (T)$. Thus increasing pressure or decreasing n (by removal of inerts) tends to drive the reaction to produce more moles of C, at a given temperature. Decreasing pressure or adding inerts promotes decomposition.

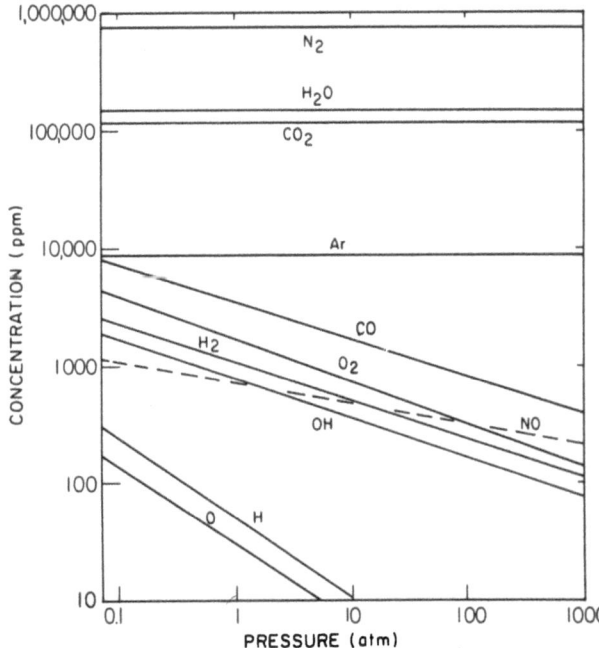

Figure 2. Effect of pressure on equilibrium composition. Stoichiometric propane/air at a temperature of 2000°K.

ADIABATIC FLAME TEMPERATURE

For processes occurring at constant volume the heat is given by

$$Q = \Delta U \quad (\text{const. } V)$$

and at constant pressure

$$Q = \Delta H \quad (\text{const. } P)$$

Adiabatic processes with either of these constraints are then simply defined as

$$\Delta U = 0 \quad (\text{adiabatic constant } V) \tag{44}$$

$$\Delta H = 0 \quad (\text{adiabatic constant } P) \tag{45}$$

For adiabatic processes these equations can be used to determine the final temperature.

If we consider a constant pressure process we have H=0 where ΔH is given by equation 33,

$$\Delta H = 0 = \sum_{i=1}^{m} n_{i_2} h_i (T_2) - \sum_{i=1}^{m} n_{i_1} h_i (T_1) \tag{46}$$

All the n_{i_1} and T_1 (i.e. the initial conditions) are known. If the final composition is known from measurements on an actual system this equation can be used to calculate the final adiabatic temperature, T_2, provided the data on enthalpy as a function of temperature are known. Most applications require computer techniques. A trial and error solution is necessary since the equation cannot be solved explicitly for T_2. For such applications curve fit equations giving the enthalpy as a power series or asymptotic series in temperature are used.

The adiabatic flame temperature can be calculated for a final equilibrium composition by simultaneously solving the equations for equilibrium composition and equation 46 for T_2. Since the equilibrium composition of the combustion products is dependent upon the temperature and the adiabatic flame temperature is dependent upon the composition, the solution of the general adiabatic flame temperature problem is complex and a sophisticated numerical technique is required. The adiabatic flame temperatures and equilibrium composition presented here, unless referenced otherwise, are the computer results following the method and using a modification of the program of Zeleznik

Table 3. Equilibrium composition of a stoichiometric mixture of hydrogen and oxygen $\phi = 1.0$, $T = 4000^{\circ}K$, $P = 1$ atm.

Species	Mole Fraction
H	.50223
H_2	.10037
H_2	.02970
O	.24974
O_2	.02851
OH	.08944

Table 4. Product composition for the combustion of heptane and air. $P = 1$ (atm), $T = 1500^{\circ}K$, varying equivalence ratios. Composition in mole fractions.

ϕ	.25	.33	.50	.67	1.00	2.00
A_r	.00971	.00911	.00900	.00889	.00868	.00686
CO_2	.03271	.04334	.06422	.08459	.12381	.03370
H_2O	.03736	.04951	.07337	.09665	.14155	.07819
N_2	.76596	.76124	.75200	.74299	.72576	.57354
NO	.00111	.00104	.00089	.00072	.00002	.00000
O_2	.15364	.13569	.10046	.06609	.00005	.00000
OH	.00005	.00006	.00006	.00006	.00001	.00000
CO	.00000	.00000	.00000	.00000	.00009	.16211
H_2	.00000	.00000	.00000	.00000	.00004	.14559
H	.00000	.00000	.00000	.00000	.00000	.00001

and Gordon (8, 9). This computer program is available to those
desiring to make such calculations and having access to the
necessary computer facilities.

The adiabatic combustion process is illustrated in Figure
3, which shows a plot of reactant and product enthalpy as a
function of temperature. The two curves represent the two terms
in equation 46. Note that the product enthalpy varies with
temperature not only due to the change in species enthalpy with
temperature but also due to the change in equilibrium product
as temperature increases.

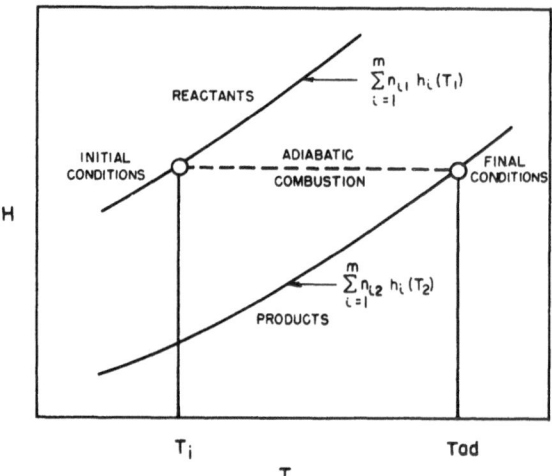

Figure 3.   Compustion reaction representation on an enthalpy-
temperature diagram.

Adiabatic flame temperatures for a number of reactants are
presented in Table 5. A range of hydrocarbon/air reactants
are seen to have similar adiabatic flame temperatures of about
2250°K at their stoichiometric mixture and a pressure of one
atmosphere.

The dependence of the adiabatic flame temperature on various
parameters represents various tradeoffs. The temperature and
product composition for n-heptane/air combustion at a constant
pressure of 5 atmospheres are given in Table 6. The flame
temperature is seen to be a maximum near $\phi = 1$ for a given pressure.
Very rich mixtures ($\phi > 1$) produce large amounts of incomplete
combustion products CO and H_2. Because of this they have a lower
flame temperature. Very lean mixtures ($\phi > 1$) have a large amount

of dilutent oxidant which absorbs the heat release from the exothermic reaction, thus lowering the final temperature. The
stoichiometric composition represents a tradeoff between these
effects and provides the maximum temperature.

Table 5. Adiabatic flame temperatures (in $^\circ$K). Gaseous reactants
at 298°K, one atmosphere pressure, stoichiometric mixtures.

Fuel	Oxidizer			
	O_2	F_2	Air	NO_2
H_2	3079	3994	2384	3029
NH_3	2845	3857	2074	2809
N_2H_4	3027	3934	2435	2992
CH_4	3054	3645	2227	2991
C_3H_8	3095	3296	2268	3029
C_8H_{18}	3108	3114	2277	3040

In the same table the maximum nitric oxide concentration is
seen to occur at an equivalence ratio of about $\phi = 0.8$. This
represents a tradeoff between the conditions which govern the
nitric oxide production. Table 4 indicates that nitric oxide
concentration increases as equivalence ratio decreases at a
given temperature. Figure 1 indicates that nitric oxide concentration increases as temperature increases at a given equivalence ratio. Thus increasing the equivalence ratio from 0 to 1
in an adiabatic process produces an effect (higher temperature)
which tends to favor nitric oxide production even though an increased equivalence ratio tends to lower nitric oxide concentrations. The maximum NO concentration at $\phi = 0.8$ is the result.

Figure 4 shows the product composition and flame temperature
as a function of pressure for constant pressure, adiabatic combustion of stoichiometric ammonia/oxygen mixtures. It is readily
seen that an increased pressure tends to increase the flame temperature by decreasing the concentrations of decomposition species
(O, H, and OH radicals). This is a direct result of the effect
of pressure on equilibrium compositions. (Increased pressure tends
to drive reactions in the direction of higher molecular weight).

Note that this also represents a tradeoff since higher temperatures
would favor the production of the high enthalpy decomposition
species.

Table 6.   Flame temperatures and product composition for the
           adiabatic combustion of heptane/air mixtures at a
           pressure of 5 atmospheres. Mole fractions for varying
           equivalence ratios. Composition in mole fractions.

ϕ	.2	.4	.6	.8	1.0	1.5
T[°K]	840	1298	1697	2046	2302	1982
Ar	.00920	.00907	.00893	.00893	.00880	.00770
CO_2	.02627	.05176	.07650	.1004	.11376	.05299
H_2O	.03002	.05915	.08734	.11414	.13747	.12580
N_2	.76939	.75786	.74601	.73391	.72004	.64100
O_2	.16511	.12194	.07918	.03753	.00403	.00000
NO	.00000	.00032	.00184	.00357	.00220	.00001
OH	.00000	.00000	.00017	.00118	.00215	.00010
CO	.00000	.00000	.00000	.00044	.00934	.11220
H_2	.00000	.00000	.00000	.00011	.00204	.0569
O	.00000	.00000	.00000	.00008	.00014	.00001
H	.00000	.00000	.00000	.00000	.00020	.00049

The calculation of the final temperature in an adiabatic
constant volume process is based on an equation analagous to (46)
using the internal energy

$$\Delta U = 0 = \sum_{i=1}^{m} n_{i_2} u_i (T_2) - \sum_{i=1}^{m} n_{i_1} u_i (T_1) \qquad (47)$$

Again this equation can be used with measured product compositions
or as part of an equilibrium calculation. Equilibrium calculations
involve an additional equation since the final pressure is not
known for the constant volume case. This pressure, however, is
simply given by the ideal gas law. Thus equilibrium, adiabatic,
constant volume combustion problems require computer solutions
with one additional unknown (and an additional equation) by

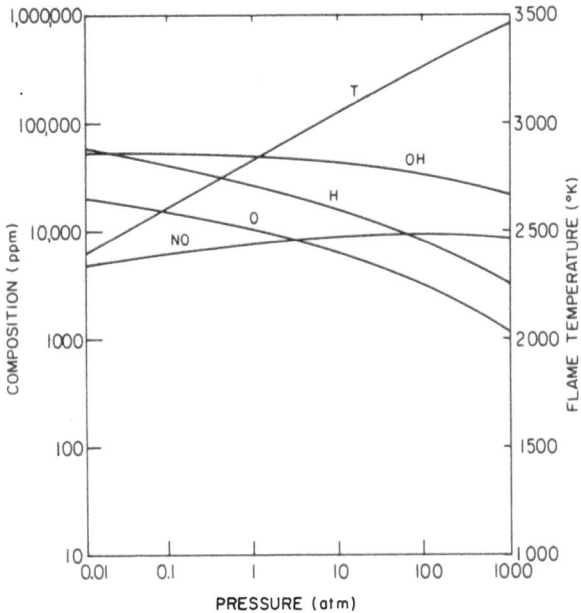

Figure 4.   Effect of combustion pressure on adiabatic flame
            temperature and product decomposition.   Gaseous
            ammonia/oxygen reactants at 298°K, ϕ = 1.0.

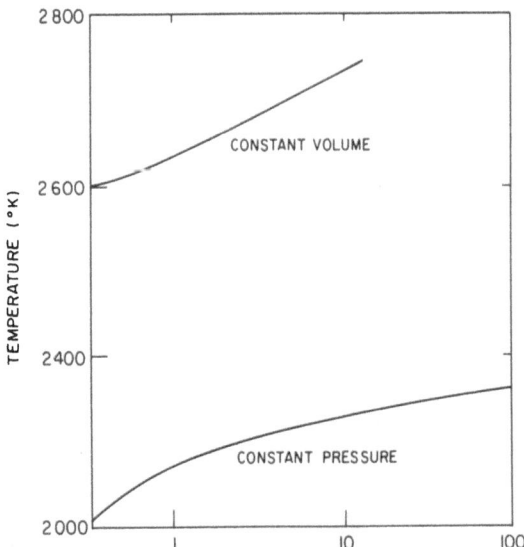

Figure 5.   Comparison of the constant volume and constant pressure
            combusion of octane and air (from data of Reference 10).

techniques similar to those used for the constant pressure case.

In a constant volume combustion there is no work done on the surroundings as in the case for the constant pressure combustion. Constant volume combustion adiabatic flame temperatures are therefore higher than the adiabatic flame temperatures for constant pressure combustion. A comparison for the combustion of octane and air is presented in Figure 5.

The adiabatic flame temperature is also affected by the initial condition of the reactants, that is, their induction enthalpies. The term induction enthalpy refers to the enthalpy of the reactants at the beginning of the combustion process, e.g., as determined by their temperature and phase. The effect of the induction enthalpy jet fuel and air is shown in Figure 6. The increase in flame temperature is roughly equal to the increase in the air induction temperature.

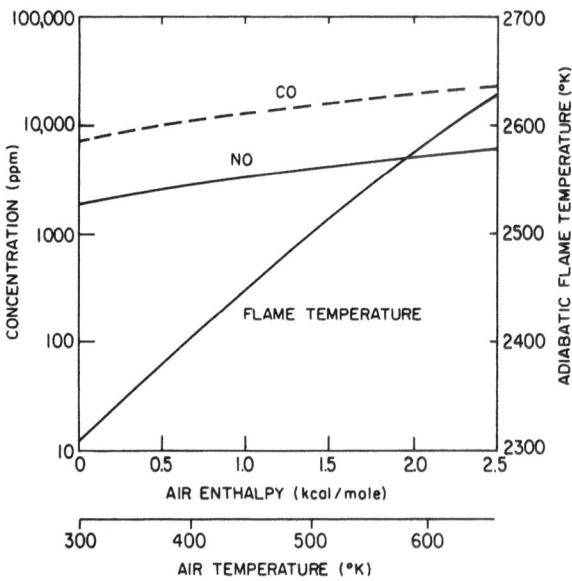

Figure 6. Effect of air induction enthalpy on adiabatic flame
           temperature and equilibrium composition. Stoichio-
           metric jet fuel and air.

SUMMARY

The foregoing considerations of combustion have been based on the assumption of the attainment of thermodynamic equilibrium in the combustion process. The model is useful and valid in predicting the energy release, flame temperature, and limiting composition of combustion product mixtures.

A consideration of the rates of combustion reactions and explanation of observed non-equilibrium combustion product combustion requires a more detailed description of the combustion process. The role of chemical kinetics and the structure of the combustion flame are important to the understanding of combustion phenomena. These topics will be considered in the succeeding sections.

An extended bibliography of books on combustion topics is included in the list of references, (11 - 34). Four periodicals devoted to combustion topics, (35 - 38) also are listed.

Symbols and nomenclature

A	Helmholtz function
C	specific heat
C_P	specific heat at constant pressure
C_V	specific heat at constant volume
d	symbol for exact differential
$đ$	symbol for inexact differential
G	Gibbs function
$\Delta G_R{}^\circ$	Gibbs free energy of reaction
H	enthalpy
$\Delta H_f{}^\circ$	enthalpy of formation with respect to standard state elements
ΔH_R	enthalpy of reaction
i	index
j	index

K_P equilibrium constant in terms of partial
                          pressures

K_n equilibrium constant in terms of number of
                          moles

m   number of reactant species, i

n_i number of moles of species i

n   total number of moles

P   total pressure

P_i partial pressure, species i

Q   heat

S   entropy

T   temperature

U   internal energy

ΔU_f° internal energy of formation

ΔU_R internal energy of reaction

V   volume

W   work

ϕ equivalence ratio

μ_i chemical potential

ν_i', ν_i'' stoichiometric coefficients

References and Bibliography on Combustion

1.   Cambel, A. B., Energy R&D and National Progress, (U. S.
        Government Printing Office, Washington, 1966).

2.   Angrist, Stanley W., Direct Energy Conversion, (Allyn and
        Bacon, Boston, 1965).

3.   Zemanski, M. W., Basic Engineering Thermodynamics, (McGraw-
        Hill, New York, 1966).

4.   Pitzer, K. S. and Brewer, L., (revision of Lewis and
      Randall), Thermodynamics, 2nd Ed., (McGraw-Hill, New
      York, 1961).

5.   JANAF Thermochemical Tables, (Clearinghouse for Federal
      Scientific and Technical Information, U. S. Department
      of Commerce, Springfield, Virginia, 1965; addendum 1966).

6.   Gordon, S., Zeleznik, F. J., and Huff, Vearl, N., A
      General method for automatic computation of equilibrium
      compositions and theoretical rocket performance of pro-
      pellants, NASA TN D-132 (October, 1959).

7.   Wilkins, R. L., Theoretical Evaluation of Chemical Pro-
      pellants, (Prentice-Hall, Englewood Cliffs, New Jersey,
      1963).

8.   Zeleznik, F. J. and Gordon, S., A general IBM 704 or
      7090 computer program for computation of chemical
      equilibrium compositions, rocket performance, and
      Chapman-Jouguet detonations, NASA TN D-1454 (October,
      1962).

9.   Gordon, S. and Zeleznik, F. J., A general IBM 704 or
      7090 compueter program for computation of chemical
      equilibrium compositions, rocket performance, and
      Chapman-Jouguet detonations, Supplement I -- Assigned
      area-ratio performance, NASA TN D-1737 (October, 1963).

10.   Steffensen, R. J., Agnew, J. T., and Olsen, R. A., Tables
      for adiabatic gas temperature and equilibrium composition
      of six hydrocarbons, Purdue University, Engineering
      Extension Series No. 122 (May, 1966).

11.   Gaydon, A. G.,Spectroscopy and Combustion Theory, 2nd Ed.,
      (Chapman and Hall, London, 1948).

12.   Jost, Wilhelm, Explosion and Combustion Processes in Gases,
      Translated by H.O.Croft, (McGraw-Hill, New York, 1946).

13.   Lewis, B., Pease, R. N., and Taylor, H. S. (Editors),
      Combustion Processes, (Princeton University Press,
      Princeton, New Jersey, 1956).

14.   Lewis, Bernard and von Elbe, Guenther, Combustion Flames
      and Explosions of Gases, 2nd Ed., (Academic Press, New
      York, 1961).

15.   Minkoff, G. J., and Tipper, C. F. H., Chemistry of
      Combustion Processes, (Butterworths, London, 1962).

16.   Mullins, B. P. and Penner, S. S., Explosions, Detonations, Flammability, and Ignition (Pergamon, New York, 1962).

17.   Penner, S. S., Chemistry Problems in Jet Propulsion, (Pergamon, New York 1957).

18.   Sokolik, A. S., Self-Ignition Flame and Detonation in Gases (IPST, Jerusalem, 1963).

19.   Spalding, D. B., Some Fundamentals of Combustion (Academic Press, New York, 1955).

20.   Strehlow, R. A., Fundamental of Combustion, (International Textbook Company, Scranton, Pennsylvania, 1968).

21.   Vulvis, Lev Abramovich, Thermal Regimes of Combustion, translated by Morris D. Friedman, (Mc-Graw-Hill, New York, 1961).

22.   Williams, F. A., Combustion Theory, (Addison-Wesley, Reading, Massachusetts, 1965).

23.   Wolfhard, H. G., Classman, I., and Green, L., Jr., (Editors), Heterogeneous Combustion, Vol. 15, Progress in Astronautics and Aeronautics, (Academic Press, London, 1964).

24.   Proceedings of the First Symposium on Combustion and the Second Symposium on Combustion, (The Combustion Institute, Pittsburgh, 1965).

25.   Third Symposium on Combustion and Flame and Explosion Phenomena, (Williams and Wilkins, Baltimore, 1949).

26.   Fourth Symposium (International) on Combustion, (Williams and Wilkins, Baltimore, 1953).

27.   Fifth Symposium (International) on Combustion, (Reinhold, New York, 1955).

28.   Sixth-Symposium (International) on Combustion, (Reinhold, New York, 1957).

29.   Seventh Symposium (International) on Combustion, (Butterworths, London, 1959).

30.   Eighth Symposium (International) on Combustion (Williams and Wilkins, Baltimore, 1962).

31.   Ninth Symposium (International) on Combustion, (Academic Press, New York, 1963).

32.   Tenth Symposium (International) on Combustion, (The
      Combustion Institute, Pittsburgh, 1965).

33.   Eleventh Symposium (International) on Combustion, (The
      Combustion Institute, Pittsburgh, 1967).

34.   Twelfth Symposium (International) on Combustion, (The
      Combustion Institute, Pittsburgh, 1969).

35.   Combustion and Flame, Beginning, Vol. 1 March, 1957
      (Journal of The Combustion Institute, Butterworths,
      London).

36.   Combustion, Explosions, and Shockwaves, (Translated,
      Fixika Goreniya i Vzyva), Beginning, Vol. 1, January,
      1965 (Faraday Press, New York).

37.   Pyrodynamics, Beginning, Vol. 1, January 1964 (Gordon
      and Breach, London).

38.   Combustion Science and Technology, Vol, 1, 1959 (Gordon
      and Breach, London), replaces Pyrodynamics.

COMBUSTION RATES

Robert F. Sawyer

Associate Professor of Mechanical Engineering

University of California, Berkeley

INTRODUCTION

The equilibrium consideration of combustion, as presented in Chapter 1, leaves many of the critical questions regarding combustion processes unanswered. Primary among these questions is that of the rate of the combustion process. Some of the most vexing problems of air pollution arising from combustion processes are those resulting from the failure to obtain equilibrium conditions. Concentration levels of hydrocarbons, carbon monoxide, and nitric oxide emitted from combustors characteristically exceed the predicted equilibrium values for the exit conditions. The rate of the combustion process determines not only the efficiency of the combustor, but, additionally, the characteristics of the combustion products.

A good description of the details of the combustion process has defied elucidation for all but the simplest combustion models. In addition to the chemical reaction involved, the effects of heat transfer, fluid mechanics, and mass transfer are likely to play an important role. Good descriptions are limited to those few combustion processes in which most of the above effects can be ignored. By a "good definition," it is implied that a model for combustion process can be formulated and analyzed which will predict the observed structure of the zone, the rate of combustion, and properties of the combustion products.

A consideration of combustion rates should yield some insight on the answers to two basic questions. First, will the combustion go to completion (equilibrium) in the time of space available? and, second, what controls the composition of the combustion products?

For real combustors the answer to the first is in most cases, <u>no</u>. As suggested previously, if the combustion processes were to go to equilibrium products, the air pollution problem would largely disappear. Although a precise answer to the second question is unlikely, some general answers are both possible and worthwhile.

CHEMICAL KINETICS

The chemical kinetics of combustion processes is concerned primarily with gas phase reactions. A notable exception is the burning of coal which occurs largely in the solid phase. Chemical kinetics focuses upon how chemical reactions occur and is divorced from the complications of heat transfer, mass transfer, and fluid mechanics which are important to combustion processes.

Although one writes a stoichiometric chemical reaction for the oxidation of methane as

$$CH_4 + 2O_2 \rightarrow CO_2 + H_2O \tag{1}$$

it is recognized that the reaction products are not limited to carbon dioxide and water but are better estimated by considerations of chemical thermodynamics. From considerations of chemical kinetics it also is recognized that oxidation of methane involves a number of steps and is of much greater complexity than would be suggested by the above equation. A reaction as written above is referred to as an "overall" or "global" reaction.

A detailed description of the chemical reactions which occur in the oxidation of methane is referred to as a reaction mechanism. One such mechanism which has been suggested for methane oxidation (1) invokes 36 reactions (represented by 18 equations describing both forward and reverse reactions).

$$CH_4 + M \rightleftharpoons CH_3 + H + M \tag{2}$$

$$CH_4 + O_2 \rightleftharpoons CH_3 + HO_2 \tag{3}$$

$$O_2 + M \rightleftharpoons 2O + M \tag{4}$$

$$CH_4 + O \rightleftharpoons CH_3 + OH \tag{5}$$

$$CH_4 + H \rightleftharpoons CH_3 + H_2 \tag{6}$$

$$CH_4 + OH \rightleftharpoons CH_3 + H_2O \tag{7}$$

$$CH_3 + O \rightleftharpoons H_2CO + H \tag{8}$$

$$CH_3 + O_2 \rightleftharpoons H_2CO + OH \tag{9}$$

$$H_2CO + O_2 \rightleftharpoons HCO + H_2O \tag{10}$$

$$HCO + OH \rightleftharpoons CO + H_2O \tag{11}$$

$$CO + OH \rightleftharpoons CO_2 + H \tag{12}$$

$$H + O_2 \rightleftharpoons O + OH \tag{13}$$

$$O + H_2 \rightleftharpoons H + OH \tag{14}$$

$$O + H_2O \rightleftharpoons 2OH \tag{15}$$

$$H + H_2O \rightleftharpoons H_2 + OH \tag{16}$$

$$H + OH + M \rightleftharpoons H_2O + M \tag{17}$$

$$CH_3 + O_2 \rightleftharpoons HCO + H_2O \tag{18}$$

$$HCO + M \rightleftharpoons H + CO + M \tag{19}$$

An important common feature of such reaction mechanisms is that in addition to the reactant and products species, a number of intermediate species are involved. These intermediates include free radicals. Although the free radicals are highly transitory, they are essential to and often control the progress of the reaction, i.e. the conversion of reactants to products. For example, the oxidation of the intermediate, carbon monoxide, to carbon dioxide depends strongly upon the presence of the hydroxyl radical (see equation 12 above). An understanding of the reaction mechanism thereby becomes useful in tracing the source of air pollutants in combustion processes.

Two approaches to the problem of the rate of a chemical reaction are invoked. For the purposes of design, the crude description provided by the overall or global reaction may be sufficient. The rate is then referred to as the global reaction rate and usually tells how rapidly reactants are converted to products. The rate of oxidation of methane as measured in a particular experimental apparatus (2) can be expressed as

$$\frac{d[CH_4]}{dt} = 10^{15.3} [CH_4]^{0.7}[O_2]^{0.8}\exp(-62500/RT) \tag{20}$$

$$[mole/cm^3]^{-.5}sec^{-1}$$

The important features are that the rate of reaction depends upon

the concentration of reactants (and sometimes products or intermediates as well) and upon the temperature at which the reaction
takes place. The dependence upon temperature is generally strong
(exponential in temperature). The dependence of the global reaction
rate upon the pressure is contained in the concentration terms. The
sum of the exponents of the concentration terms is the order of the
reaction. The rate of a first order reaction is independent of
pressure (the rate of the fractional disappearance of reactants
does not change with pressure at a fixed temperature.) A good
global raction rate measurement should be independent of the
apparatus in which it was measured. In practice this is often not
true suggesting that the effects of heat and mass transfer and
fluid mechanics may not have been eliminated. The above rate
equation could be written as

$$\frac{d[CH_4]}{dt} = k[CH_4]^{0.7}[O_2]^{0.8} \tag{21}$$

where, $k = 10^{15.3}exp(-62500/RT)$ is known as the Arrhenius rate constant.

A more recently popular approach to the measurement and prediction of reaction rates focuses upon the reaction mechanism. If
the rates of the individual steps were known, and if the reaction
mechanism were accurate, it should be possible to determine the
overall reaction rate from the steps. This approach, since it is
more general than the global measurements, is perhaps more scientifi-
cally satisfying. As better measurements of individual reaction rate
constants, k, become available, the utility of the reaction mechanism approach will increase.

For an individual reaction written as

$$\sum_{i=1}^{n} \nu_i' A_i \rightarrow \sum_{i=1}^{n} \nu_i' A_i \tag{22}$$

the rate may be expressed as

$$\frac{d[A_j]}{dt} = k(\nu_j'' - \nu_j') \prod_{i=1}^{n} [A_i]^{\nu_i'} \tag{23}$$

For a complex reaction mechanism, the rate of disappearance of a
species, A_j may be written in general form as

$$\frac{d[A_j]}{dt} = \sum_{k=1}^{m} \left\{ k_k (\nu_{jk}'' - \nu_{jk}') \prod_{i=1}^{n} [A_i]^{\nu_{ik}'} \right\} \tag{24}$$

If similar expressions are written for all species to include the contributions of all reactions, a set of non-linear, ordinary, differential equations results. The solution to these equations gives the compositions of all species as a function of time. The rate of disappearance of the reactant corresponds to the global reaction rate described previously. The digital computer makes such solutions feasible. (See, for example, reference 3.)

Because of the important role played by chemical kinetics in air pollution, an understanding of kinetics is essential to the understanding of air pollution. Additional ideas from chemical kinetics will be presented in following consideration of the photochemistry of air pollution, the description of particular sources of air pollution, and the chemical analysis of source pollutants.

FLAMMABILITY LIMITS AND IGNITION TEMPERATURE

The composition of combustible mixtures and the temperature at which combustion may be initiated are particular properties which depend upon rate processes. If the rate of reaction is sufficient to just balance the rate at which heat is lost, combustion is sustained. This condition is dependent upon the composition, temperature, pressure, container geometry, container material, and other reactant and system properties. If all conditions except the composition are held constant, limits upon the composition (equivalence ratio) are noted, beyond which combustion of the propagation of a flame is not self-sustained. These limits are referred to as the flammability limits. Paraffin hydrocarbon flammability limits are recorded in Table 1.

The autoignition temperature refers to the lowest temperature at which ignition can occur. Measurements usually are made in large containers to minimize wall effects. Autoignition temperatures for paraffin hydrocarbons are listed in Table 2.

FLAMES

A flame is identifiable with many combustion processes, and forms one focus for consideration of combustion rates. A flame is described as a combustion reaction which can propagate subsonically in space. Detonations (supersonic propagation) and volume reactions (such as stirred reactors) are other combustion processes.

The concept of a moving flame, or a stationary flame in a flowing gas, implies that there exists an unburned and burned region separated by a zone whose dimension necessarily is small in comparison to the system. As a consequence, the concept of a flame zone arises. Other equivalent terms are flame front, combustion wave, combustion zone, deflagaration wave, etc. The propagation of the flame is a result of a feedback from the reacted region to the unreacted region. This feedback is one of

Table 1

Flammability limits of paraffin hydrocarbons in air at 25°C (4).

hydrocarbon	formula	lower limit mole fraction hydrocarbon	upper limit mole fraction hydrocarbon
methane	CH_4	.050	.150
ethane	C_2H_6	.030	.124
propane	C_3H_8	.021	.095
n–butane	C_4H_{10}	.018	.084
n–pentane	C_5H_{12}	.014	.078
n–hexane	C_6H_{14}	.012	.074
n–heptane	C_7H_{16}	.0105	.067
n–octane	C_8H_{18}	.0095	–
n–nonane	C_9H_{20}	.0085	–
n–decane	$C_{10}H_{22}$	.0075	.056
n–undecane	$C_{11}H_{24}$	.0068	–
n–dodecane	$C_{12}H_{26}$	.0060	–
n–tridecane	$C_{13}H_{28}$	.0055	–
n–tetradecane	$C_{14}H_{30}$	.0050	–
n–pentadecane	$C_{15}H_{32}$	.0046	–
n–hexadecane	$C_{16}H_{34}$	.0043	–

Table 2

Autoignition temperatures of paraffin
hydrocarbons at atmospheric pressure   (4)

hydrocarbon	formula	autoignition in air	temperature in oxygen
methane	CH_4	537°C	-
ethane	C_2H_6	515	506°C
propane	C_3H_8	466	-
n-butane	C_4H_{10}	405	283
iso-butane	C_4H_{10}	462	319
n-pentane	C_5H_{12}	258	258
n-hexane	C_6H_{14}	223	225
n-heptane	C_7H_{16}	223	209
n-octane	C_8H_{18}	220	208
n-nonane	C_9H_{20}	206	-
n-decane	$C_{10}H_{22}$	208	202
n-dodecane	$C_{12}H_{26}$	204	-
n-hexadecane	$C_{16}H_{34}$	205	-

transfer of energy or active species and is brought about by the
mechanism of conduction, radiation, or diffusion, or a combination
of these mechanisms. Note that the steady state propagation of a
flame is a unique case for which the feedback is just sufficient
to maintain the combustion reaction. If the feedback is less, the
flame will cease to propagate, i.e., it will be extinguished. If
the feedback is more than is required to sustain the steady state
flame, the rate will increase and a transition from a flame to a
detonation will occur.

Important characteristics of flames, to summarize, are that
they:

(1) involve exothermic chemical reactions
(2) possess the ability to propagate through space
(3) usually involve a fuel/oxidizer system (but also in-
     clude such processes as exothermic decompositions and
     recombinations)
(4) are usually gas phase
(5) are usually hot
(6) have characteristic times ranging from seconds to milli-
     microseconds.

The most important processes relating to flames are those of
thermal conduction, molecular diffusion, and chemical reaction.
The subject of chemical reaction has been treated briefly in the
preceding section on chemical kinetics. A detailed consideration
of the many facets of heat and mass transfer is beyond the scope
of this presentation. The basic concepts, however, that the rate
of heat transfer is proportional to a temperature gradient and
that the rate of diffusion of mass is proportional to a concentration gradient, are essential to consideration of flames.

Flames may be classified according to the following characteristics:

(1) homogeneous vs. heterogeneous (in phase)
(2) premixed vs. unmixed
(3) laminar vs. turbulent
(4) monofuel vs. fuel/oxidizer
(5) steady vs. non-steady

Three types of flames are considered here, primarily for the information which may be derived regarding how system parameters
(temperature, pressure, fluid properties, etc.) affect the combustion rate. These are, in order, laminar premixed flames,
laminar diffusion flames, and premixed turbulent flames.

Laminar premixed flames

This category of flames is limited, for the purpose of this presentation, to those to which the following descriptors apply: laminar, premixed, deflagration, fuel/oxidizer, homogeneous, gas phase, steady, one dimensional, propagating, adiabatic. The problem to be solved is (a) to determine the structure, i.e., the composition and temperature as a function of distance and (b) to determine the propagation velocity. The Bunsen flame is a typical laminar premixed flame.

Schematically, this flame may be represented according to the model of **Figure 1.** An analytical description requires statements expressing, through the flame zone,

   (1) conservation of mass
   (2) conservation of species
   (3) conservation of momentum
   (4) conservation of energy

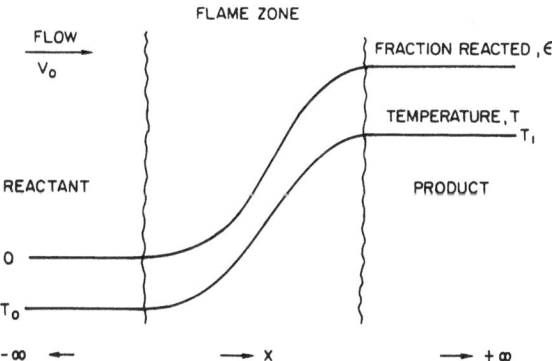

Figure 1. Flame zone model for a laminar premixed flame.

For the general case, a large number of differential equations result (there is a conservation of species equation for each species). By assuming that viscous effects are negligible, kinetic energy is negligible, and that all species have the same, constant specific heat, one may obtain a considerably simplified set of describing equations.

$$\rho v = \dot{m} = \text{constant} \tag{25}$$

$$P = \text{constant} \tag{26}$$

$$\dot{m}[C_p T - (1 - \varepsilon)\Delta H_r] - \lambda \frac{DT}{dx} = E = \text{constant} \tag{27}$$

With the elimination of the role of diffusion, these equations describe a "thermal propagation model." That is, the propagation of the flame results from a feedback of thermal energy from the reacted zone to the unreacted zone. The boundary conditions for the unreacted, "upstream" region are:

at $x = -\infty$:

$$T = T_o$$

$$\frac{dT}{dx} = 0$$

$$\varepsilon = 0$$

And, for the reacted, "downstream," region:

at $x = +\infty$:

$$T = T_1$$

$$\frac{dT}{dx} = 0$$

$$\varepsilon = 1$$

It is desired to solve equations (1), (2), and (3) to give the temperature as a function of the pertinent parameters. That is, the solution is to take the form:

$$T = T(x, \lambda, C_p, \dot{m}, T_1, T_o) \tag{28}$$

The actual solution of even these "simple" equations requires, however, specification of the relation of the fraction reacted, ε, to the distance, x, or temperature, T. Without actually solving the equations, one may determine from their form that a characteristic solution (eigenvalue) exists and yields a functional dependence for the flame speed, v_o, of

$$v_o = N \sqrt{\frac{\lambda}{\rho_o C_P} \overline{\frac{d\varepsilon}{dt}}} \qquad (29)$$

where:

$$\frac{\lambda}{\rho_o C_P} = \text{thermal diffusivity}$$

$$\frac{d\varepsilon}{dt} = \text{reaction rate}$$

$$N = \text{experimental value}$$

With appropriate additional assumptions relating the fraction reacted to the temperature or distance, one can obtain solutions for the propagation velocity of the laminar flame. Most results have the common feature that the flame velocity, i.e., the combustion rate, is proportional to the square root of both the thermal diffusivity and the reaction rate.

Table 3

Experimentally observed characteristics of laminar premixed flames. Fuel burning in air (6).

fuel	equivalence ratio	flame velocity cm/sec	adiabatic flame temperature ^{o}K
C_2H_2	1.33	163	
C_6H_6	1.08	47	2306
CO	1.70	45	
H_2	1.70	306	
CH_4	1.06	39	2236
C_2H_6	1.12	46	2244
$n-C_4H_{10}$	1.13	44	2256
$n-C_6H_{14}$	1.17	45	2238
$n-C_7H_{16}$	1.22	45	2238
jet fuel, JP-4	1.07	40	

The "complete" describing equations, including diffusion of mass, may be formulated and solved by numerical means (5). Such sophisticated models allow both thermal conduction and mass diffusion of active species to be considered in the flame propagation analysis.

The most important single result of the laminar premixed flame theories is the resulting dependence of the flame velocity upon the reaction rate.

$$v_o \propto \sqrt{\frac{\overline{d\varepsilon}}{dt}} \qquad (30)$$

This relation allows the determination of the reaction rates from simple flame rate measurements.

Experimentally observed flame velocities and temperatures of premixed laminar flames of fuels burning in air are presented in Table 3-3. The values range from 40 to 300 (cm/sec). The effect of equivalence ratio on flame velocity is shown for some hydrocarbon/oxygen flames in Figure 2. The maximum flame velocity is observed near the stoichiometric mixture where temperature also is a maximum.

Laminar diffusion flames and droplet burning

In diffusion flames the reactants are not premixed. Chemical reactions necessarily are fast compared to the rate of mass transfer by diffusion and the rate of energy transfer by conduction. If such were not the case, mixing could occur before reaction and the flame would be premixed. Examples of diffusion flames are a candle burning in air and the burning of a liquid fuel droplet in an oxidizing atmosphere. The flow field in which the reaction takes place may be laminar, as for the local flame region in droplet burning, or turbulent, as for most industrial combustors and internal combustion engines.

Droplet burning may be analyzed according to the model presented in Figure 3. The location and width of the flame zone is determined by the transport properties (diffusion and thermal conductivity) and chemical kinetics. As for the laminar premixed flame, a number of assumptions and simplifications are necessary to make analysis tractable. Typical assumptions invoked are: spherical symmetry, constant pressure, fuel source at the liquid drop surface, oxidizer source at infinity, no radiation heat transfer, no dynamic effects on the droplet, equal diffusion of mass and heat.

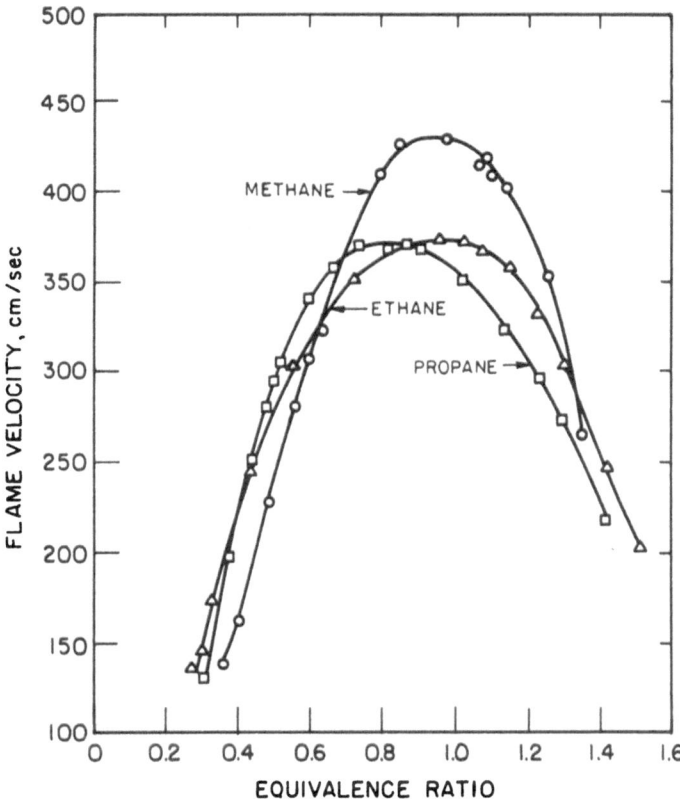

Figure 2. Flame velocity of paraffin-oxygen mixtures at one
            atmosphere pressure and room temperature. (4)

The describing equations again are expressions of conservation
of mass, conservation of species, conservation of momentum and
conservation of energy. The form of the equations indicates two
eigenvalues, the liquid drop surface temperature usually is taken
to be the boiling temperature of the liquid. If a thick reaction
zone is to be considered, a trial and error numerical solution is
required. By invoking a "collapsed reaction zone," one can write
somewhat simplified describing equations for each zone. A thin
reaction zone assumption is valid for those cases in which the
chemical reaction is very fast compared to the diffusion rates.
Another consequence of this model is that the reaction takes place
at the stoichiometric mixture, as all fuel and oxidizer diffusing
to the flame zone must be consumed. The resulting solution yields
three interesting predictions in general agreement with experimental
observations. First, the mass burning rate of the droplet is

proportional to the droplet radius to the first power and not to
the droplet surface area, or radius to the second power, as one
might expect. Second, the burning rate is independent of pressure.
And, finally, the flame approaches the droplet as burning occurs.

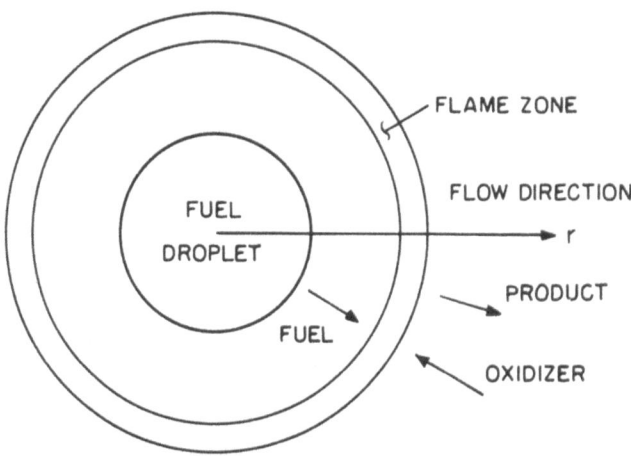

Figure 3. Laminar diffusion flame model for a fuel droplet burning in an oxidizing atmosphere.

Although droplet flames can be formed with either the fuel or
the oxidizer in the liquid state, it is typically the fuel that is
the liquid and the oxidizer that is the gas. The gross characteristics of such liquid-gas flames depend strongly upon the disper-
sion of the liquid in the gas. If the liquid is dispersed into
very fine droplets, evaporation may take place completely before
combusion and the flame takes on the character of premixed flame.
Typically such will occur if the droplet size falls below about
30 microns. If the droplets are about 30 microns and 1000 microns
(1 millimeter) in diameter, the liquid droplets burn as sperical
diffusion flames. Oil burners and turbojet combustors typically
produce burning of this type.

The burning rate generally is fixed by the evaporation rate
at the droplet surface. The burning rate may be evaluated by
equating the heat of vaporation to the heat conducted from the hot
product gas.

$$\dot{m}\Delta H_L = v_0 \rho_0 \lambda H_L = \lambda_0 \left(\frac{dT}{dr}\right)_0 \tag{31}$$

Observed burning velocities are of the order of 100 cm/sec which are comparable to premixed laminar flames.

As the "droplets" become greater than about a centimeter in diameter, the fuel geometry is better characterized as a pool. In pool burning the geometry and size restrict the availability of the oxidizer so that hydrocarbons burn with flames heavily loaded with soot. The propagation mechanism in pool burning generally is by radiation. Since sooty flames are good radiators, the burning rate approaches a value given by,

$$\dot{m} \Delta H_L = \varepsilon \sigma T_f^4 \tag{32}$$

The characteristics of the various liquid-gas flames are summarized in Figure 4.

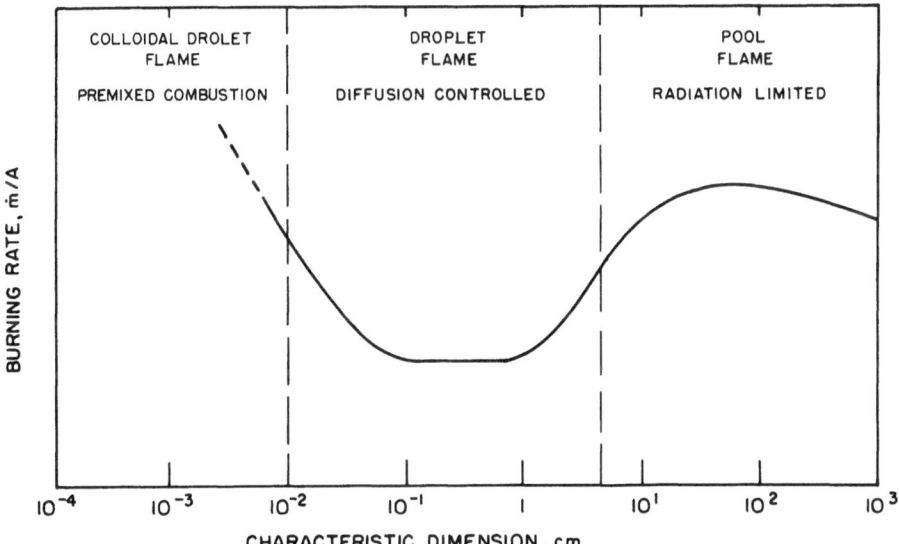

Figure 4. Effect of droplet size on the burning characteristics of liquid fuels, after (2).

Premixed turbulent flames

In the premixed turbulent flame, the complexity of chemical reaction, fluid mechanics, and mass transfer has prevented the

development of a useful analytical model. Since experimental observations show no well defined flame zone, measurement of a burning velocity is difficult. Flame velocities are of the order of
300 to 10,000 cm/sec, or considerably greater than for premixed
laminar flames. Flame velocities show some dependence on the degree of turbulence--increasing turbulence increases the flame
velocity.

The combustion processes of concern to air pollution probably
involve a combination of droplet burning (laminar diffusion flame),
turbulent diffusion, and premixed turbulent flames in the same
combustor.

An excellent treatise on flames containing the results of many
experimental flame studies is the book by Fristom and Westenberg
(6).

STIRRED REACTORS

In some combustors the reaction is distributed over a volume
rather than localized in the narrow surface of a flame. Such
reactors may be conveniently idealized as being perfectly stirred.
In the perfectly-stirred reactor (PSR) reactants instantaneously
mix with reaction products in a homogeneous region. The combustion products of PSR necessarily contain a fraction of unreacted
fuel and oxidizer. This model has proven useful in dealing with
gas turbine and furnace combustion.

The fraction of the incoming mixture which reacts, ε , is related to the reaction rate, reactant flow rate, and reactor volume
by,

$$\frac{d[R]}{dt} = \frac{\dot{N}_R \varepsilon}{V} \tag{33}$$

A second equation for the reaction rate is the first order
form of the Arrhenius expression

$$\frac{d[R]}{dt} = k[R] \tag{34}$$

These two equations may be combined to give the fraction reacted

$$\varepsilon = kV \frac{[R]}{\dot{N}_R} \tag{35}$$

The following relations

$$[R] = (1 - \varepsilon)[R]_o \tag{36}$$

$$t_s = \frac{V[R]_o}{\dot{N}_R} \qquad (\text{mean stay time}) \tag{37}$$

may be used to give

$$\varepsilon = k(1 - \varepsilon)t_s \tag{38}$$

If a characteristic reaction time is defined as $t_k = k^{-1}$, the fraction reacted may be expressed by

$$\varepsilon = \frac{t_s/t_k}{1 + (t_s/t_k)} \tag{39}$$

The fraction reacted is seen to depend on the ratio of stay time to characteristic reaction time. For a fixed characteristic reaction time, the fraction reacted increases with stay time.

Stable operating conditions may be investigated by considering the rate of energy release in the reactor and the rate of energy loss from the reactor. In general, three intersections of the energy rate curves may be identified, **Figure 5**.

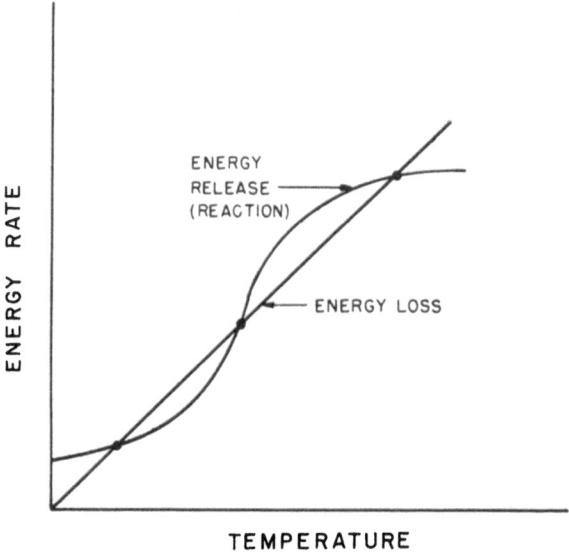

Figure 5. Stationary states of a perfectly stirred reactor.

The upper and lower intersections define stable operating points.
The central intersection is "unstable". For example, referring
to the upper intersection, if the temperature increases, the rate
of energy loss will exceed the rate of energy release and the
temperature will drop. For the central intersection, however,
any change in temperature will cause the reactor to shift to
either the upper or lower intersection.

More detailed analyses and descriptions of stirred reactors
are to be found in a number of references; for example, see (7)
and (8).

DIFFUSION AND HEAT TRANSFER IN COMBUSTION

The characteristics of combustion in general and the combustion rate in particular depend upon the rates of mass transfer,
of heat transfer, and of chemical reaction. The reaction rate has
been treated previously. In laminar flow, transfer of heat and
mass are given, respectively by

$$\dot{q} = -\lambda \frac{dT}{dx} \tag{40}$$

$$\dot{m}_i = -D_i \frac{d\rho_i}{dx} \tag{41}$$

Prediction of transport properties, in particular the thermal conductivity, λ, and the mass diffusion coefficient, D_i, for the mixtures and temperatures characteristic of combustion processes is
difficult. The lack of such information makes estimation of combustion rates in turn difficult.

Diffusion coefficients for binary mixtures have been measured.
Values are reported for some pairs of gases common to combustion
in Table 4. A strong dependence on mixture and temperature is
noted.

Table 4

Experimental diffusion coefficients. Trace amounts
of the first gas in the second. D in (cm^2/sec).(6)

T $°K$	CH_4-O_2	O_2-CO_2	H_2-O_2
300	.23	.16	.82
500	.58	.42	2.09
1000	1.95	1.43	

Thermal conductivities for some pure gases are presented in Table 5. Once again, a strong dependence on species and temperature is evident. For mixtures of gases, the selection of effective diffusion coefficients and conductivities is difficult. Often the properties of the dominant species are taken to represent the mixture.

Table 5

Experimental gas thermal conductivities
λ , in (cal/cm sec $^{\circ}$K) (6)

$^{\circ}_K T$	N_2	H_2	O_2
300	6.1	43.4	
500	9.7	65.0	8.5
1000	15.7		23.4

The effect of turbulence on the combustion rate is related directly to the much higher diffusivities of turbulent flow over laminar flow. It is a good approximation to use equations (40) and (41) for turbulent flow through the expediency of substituting a turbulent eddy diffusivity for coefficients appearing in these equations. The level of turbulence is described in terms of its intensity and scale. The intensity of the turbulence, u', is determined from the departure of the local velocity, u, from its mean value, u_o

$$u' = \{\overline{(u - u_o)^2}\}^{1/2} \tag{42}$$

The scale of the turbulence is a measure of the size of a turbulent eddy, l_2. For isotropic turbulence, the scale of the turbulence in terms of the mean distance traveled by an eddy before it loses its identity, l_1, is the same as the size of the eddy, i.e., $l_1 = l_2$. The turbulent diffusivity,

$$\varepsilon \simeq l_1 u' \tag{43}$$

These concepts apply to the diffusive transport of mass, momentum, and heat transfer in turbulent flows.

FORMATION OF AIR POLLUTANTS IN COMBUSTION PROCESSES

Although it is difficult to generalize results obtained from a particular combustor design, it is useful to look at a "real" combustor in which the effects of chemical kinetics, heat transfer,

fluid mechanics, and mass transfer all play a role. Some experimental observations of a laboratory combustor are discussed as an
example of conditions controlling the exhaust composition.

A small laboratory "can-type" combustor was studied in which
a liquid hydrocarbon fuel was burned with air. The fuel is injected axially through a single nozzle. The air enters through a
number of holes in a cylindrical surrounding "can." For purposes
of description it is convenient to divide the combustor into two
parts, a primary zone and a secondary zone, Figure 6. In the
primary zone fuel and air are mixed and partially burned at near
stoichiometric conditions. The approximation of a stirred reactor can be applied to the primary zone. In reality a lack of
homogeneity provides both locally fuel rich and fuel lean regions.
Stability is maintained through recirculation in the primary zone.
In the secondary zone additional air is provided and the reaction
is completed. Further dilution occurs in the secondary zone to
reduce the gas temperature.

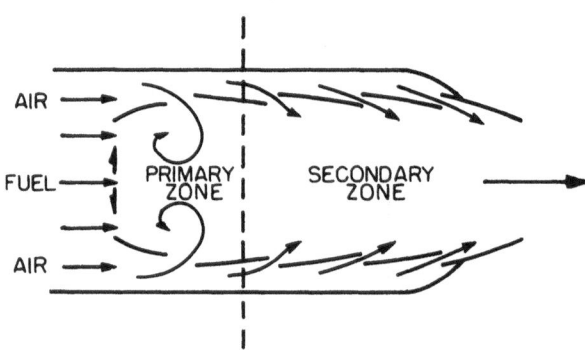

Figure 6. Laboratory combustor, one dimensional model.

Consistency with this simple model is noted from centerline
measurements of local mixture ratio, temperature, and fraction
reacted, Figure 7. The boundary separating primary and secondary zones has been selected arbitrarily at the point of stoichiometric mixture and maximum temperature.

In a very general sense, air pollution emission characteristics of most combustors are traceable to the failure of the
exhaust gas to attain chemical equilibrium. If the combustion
products were at chemical equilibrium at the exhaust conditions,
one would expect to find negligible concentrations of most of the
air pollutants, in particular, carbon monoxide, hydrocarbons, and
nitric oxide. Identifying the processes which fix the exhaust
composition is a matter first of identifying where in the flow
history of the exhaust gas conditions first arise which prevent

maintaining of chemical equilibrium. Having made this identification, one can speculate upon the chemical and/or physical processes which have caused a departure from equilibrium.

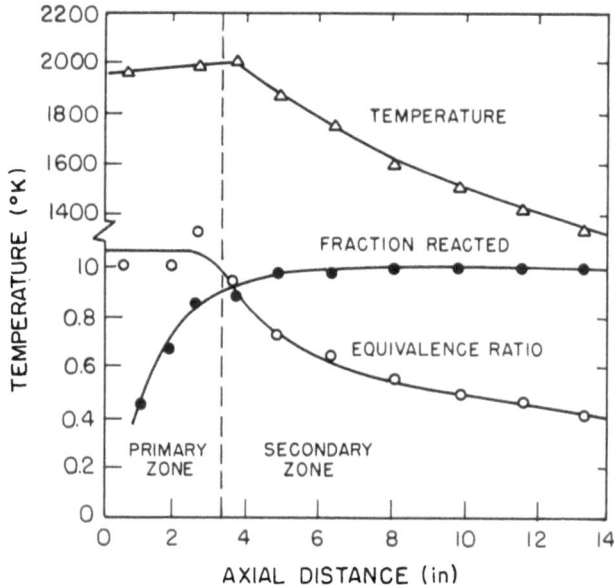

Figure 7.   Laboratory combustor, centerline reaction profile.

Carbon Monoxide

The centerline composition of carbon monoxide in the laboratory combustor is shown in Figure 8. Although these data are
for a particular set of operating conditions and apply to a sing-
ular flow path, it is proposed that it is not unreasonable to
assume that similar conditions are common to other flow paths,
at least in a qualitative sense. The equilibrium composition of
carbon monoxide calculated for the measured local equivalence
ratio also is plotted. The important general features are that
the carbon monoxide concentration in the primary zone is approximately at the equilibrium level but that an increasing departure
from equilibrium occurs as the flow passes through the secondary
zone. Throughout the secondary zone the carbon monoxide level
drops through chemical reaction to form carbon dioxide. The
carbon monoxide profile is fixed by the chemical kinetics of its
reaction which in turn is controlled primarily by the temperature
profile (really the temperature-time history) in the secondary
zone. This phenomenon is sometimes referred to as "thermal

quenching" and is not unlike problems encountered in rocket
nozzle flows and in piston engine expansion processes. Thermal
quenching will be most rapid for high air/fuel mixtures (low
equivalence ratio). It is interesting to note that in the experiment cited the cooling rate is not so great as to "freeze" the
level of carbon monoxide. Although chemical equilibrium is not
maintained, significant chemical reaction in which carbon monoxide
is consumed does occur.

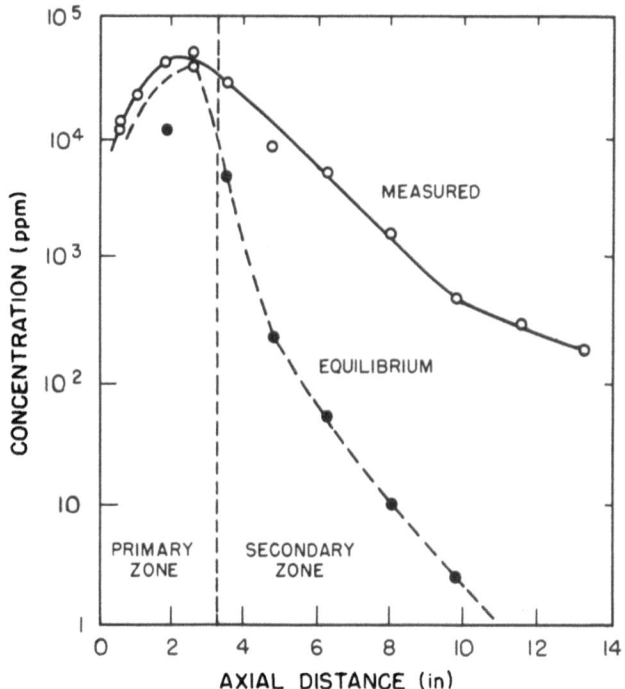

Figure 8. Laboratory combustor, centerline carbon monoxide.

Hydrocarbons

While similar in many respects to the behavior of carbon
monoxide, the hydrocarbon concentration profile is basically
different, Figure 9. The concentration of hydrocarbon in the
primary zone is not related to the equilibrium concentration,
which is predicted to be negligible. It depends, rather, upon
the fraction reacted attained in the primary zone. The concept
of a stirred reactor model for the primary zone is in good agree-

ment with this idea. For the experiment cited hydrocarbon reaction occurs rapidly in the initial region of the secondary zone
but at some point reaction ceases. As for the carbon monoxide,
attainment of chemical equilibrium, the effect of thermal quenching is much more severe and hydrocarbon concentration is "frozen."

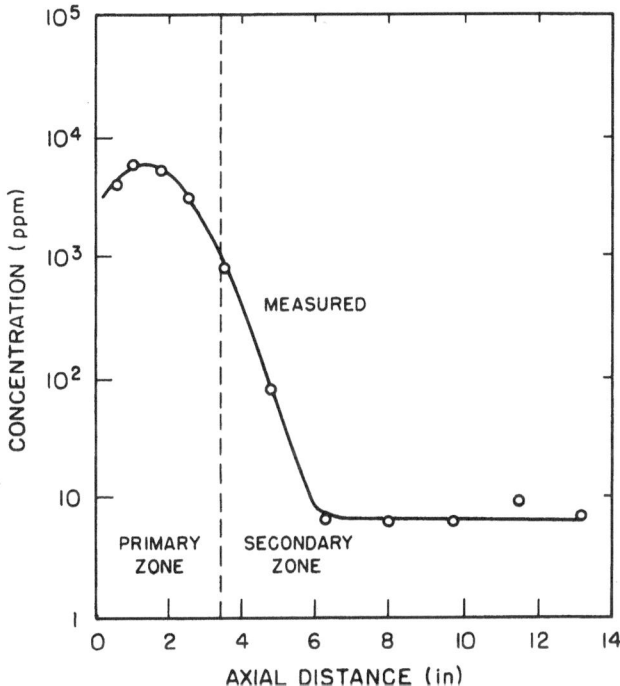

Figure 9. Laboratory combustor, centerline hydrocarbons (hexane
equivalent).

Nitric Oxide

Laboratory experiments indicate that the processes fixing
the level of nitric oxide are quite different from those for
carbon monoxide or hydrocarbons. Combustor measurements, Figure
10, show that, in addition to thermal quenching, formation
processes are important. The concentration of nitric oxide at
the boundary between primary and secondary zones is not related
to the chemical equilibrium concentration. The kinetics of the
formation of nitric oxide and, therefore, the temperature, composition, and time characteristics of the primary zone, fix the

concentration at the zonal interface. In the secondary zone
thermal quenching occurs. The slow decrease in nitric oxide
concentration is the result of dilution. The processes fixing
the nitric oxide level are much more complex than was the case
for carbon monoxide and hydrocarbons. The single most important
factor is the maximum temperature reached in the combustor.

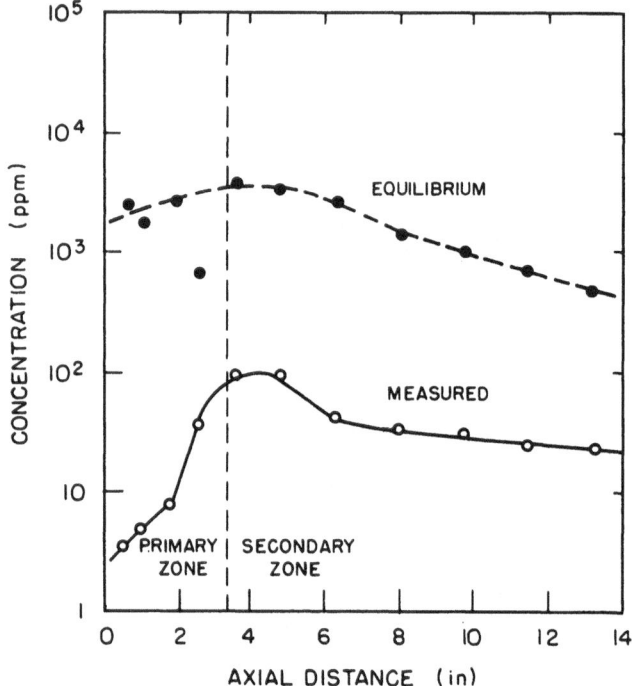

Figure 10. Laboratory combustor, centerline nitric oxide.

SUMMARY

A number of parameters have been identified which affect combustion rates. Included in such a list are the following:

(1) pressure
(2) temperature
(3) thermal diffusivity
(4) molecular diffusivity
(5) chemical reaction rate
(6) turbulence level

Some general conclusions may be made regarding the effect of each
of those parameters upon the combustion rate.

The principle effect of the pressure is to change the thickness of the flame zone or loading of a stirred reactor. Typically,
the flame thickness is inversely proportional to the pressure.
As combustion pressure is increased, combustion times decrease.
High energy density combustors are necessarily high pressure combustors.

Temperature affects the combustion rate primarily through
the exponential dependence of chemical reaction rate upon temperature. Strong variations of transport properties with temperature
introduce additional dependence of combustion rate upon temperature. Preheating of reactants provides a means of accelerating
combustion rates.

The effects of the transport properties, thermal diffusivity
and molecular diffusivity are most evident in comparing the combustion of hydrogen, which has very high diffusivities, with
most other fuels. Transport properties generally are not available as controllable system parameters which may be adjusted for
their effect on combustion rate.

In diffusion flames, chemical reaction rate does not control
the combustion rate. In premixed flames, however, chemical reaction rate plays a primary role in determining the combustion
rate. The product composition in both diffusion and premixed
flames is determined by the chemical reaction rates. Although
combustors generally are designed with stay times sufficient to
provide nearly complete energy release, there is no assurance
that this time is also sufficient to provide attainment of equilibrium in the combustion products.

High levels of turbulence greatly accelerate combustion rates.
Industrial combustors are designed to operate under turbulent conditions. The burning rate of a droplet in a turbulent flow may,
however, be controlled by the diffusion rates in the local laminar
flame zone surrounding the individual droplet.

The bridge between the disciplines of physical chemistry,
high temperature thermodynamics, and fluid mechanics and the
design of combustors necessarily requires an understanding of the
many interactions controlling combustion rates. Simplified flame
and reactor models provide such a bridge in a few idealized cases.
The utility of such models is likely to fall more immediately in
the increased understanding of observed characteristics of combustion processes, including the emission of air pollutants, than in
the original design of combustors.

Symbols and nomenclature

A	chemical species
C_P	specific heat at constant pressure
D	diffusion coefficient
ΔH_L	enthalpy of evaporation
ΔH_R	enthalpy of reaction
E	activation energy in Arrhenius expression
E	total energy
k	rate constant
l_1, l_2	turbulence scale
$\dot{m}$	mass flow rate, mass combustion rate
N	constant
$\dot{N}_R$	reactant flow rate
P	pressure
$\dot{q}$	heat transfer rate
r	radial distance
R	universal gas constant
$[R]$	reactant concentration
t	time
t_k	characteristic reaction time
t_s	stay time, mean residence time
T	temperature
T_f	flame temperature
u	flow velocity

u_o	mean flow velocity
u'	turbulent intensity
v	flow velocity
v_o	flame velocity
V	volume
x	distance
ε	emissivity
ε	eddy diffusivity
ε	fraction reacted
λ	thermal conductivity
ν	stoichiometric coefficient
ρ	density
σ	Stefan-Boltzmann constant

Notation and subscripts

[ ]	concentration
i	species index
j	species of interest
k	reaction index
m	number of reactions
n	number of species
o	unburned
1	burned

References

1. Seery, D. J. and Bowman, C. T.,"An Experimental and analytical study of methane oxidation behind shock waves," United Aircraft Research Laboratories, private communication.

2. Dryer, F. and Glassman. R., "Overall reaction of paraffin hydro-carbons," 10th AFOSR Contractors' Meeting on kinetics of energy conversion, Berkeley, California (4-5 September 1969).

3. Purgalis, P. and Sawyer, R. F., "General formulation and numer-ical solution of reaction rate problems," College of Engineering, University of California, Berkeley, Report No. TS-68-4 (June 1968).

4. Zabetakis, M. G., "Flammability characteristics of combustible gases and vapors," Bureau of Mines Bulletin 627, United States Department of the Interior (1965).

5. Von Karman, Theodore, "Structure and propagation of laminar flames," Sixth Symposium (International) on Combustion, pp. 1-11, (Reinhold, New York, 1957).

6. Fristrom, R. M. and Westenberg, A. A., Flame Structure, (McGraw-Hill, New York, 1965).

7. Vulis, L. A., Thermal Regimes of Combustion, (M. D. Friedman, translator) McGraw-Hill Book Co., New York, 1961.

8. Frank-Kamenetskii, D. A., Diffusion and Heat Exchange in Chemi-cal Kinetics, (N. Thon, translator) Princeton University Press, Princeton, N. J., 1955.

Note: An extensive list of references on combustion, including com-bustion rates, appears at the end of the first paper.

SOME FUNDAMENTAL ASPECTS OF PHOTOCHEMISTRY
AND PHOTOCHEMICAL AIR POLLUTION

James N. Pitts, Jr.

Professor of Chemistry

University of California, Riverside

I. INTRODUCTION

The purpose of the following notes is to present in general
form some principles of photochemistry and discuss some examples of
photochemistry and discuss some examples of their applications to
the problem of photochemical air pollution. Literature references
are to be found at the end of these notes.

II. FUNDAMENTALS OF PHOTOCHEMISTRY

The fundamental difference between the usual thermal reactions
and photochemical processes is in the manner in which the energy
initially required to break chemical bonds is acquired. In pyroly-
ses, energy in excess of the activation energy for reaction is obtained through a series of random collisions with other energetic
molecules in the system. In photolyses, the molecule becomes
photoactivated by absorbing a quantum of radiation of frequency ν.
If sufficient energy is provided, the excited molecules can then
dissociate, rearrange or react with another molecule. It is of
great importance that the energy of the photon absorbed by the
molecule is given exactly by the equation

$$E = h\nu \qquad\qquad (1)$$

where E is in ergs, h is Planck's constant, and ν is the frequency
in sec^{-1}. Thus, in order to achieve reaction the researcher in
pyrolytic processes can be said to use a type of shotgun technique
in which energies of excitation are distributed through a wide
range. On the other hand, a photochemist employing monochromatic
light of frequency ν is by analogy using a rifle that delivers
quanta of known energy $h\nu$.

A quantity often used by photochemists is a mole of quanta. This refers to Avogadro's number 6.02 x 10^{23}, of quanta, and is called an "Einstein" of radiation. By using the appropriate conversion factors, one can derive from equation (1) the expression

$$E = \frac{2.858 \times 10^5 \text{ kcal. Einstein}^{-1}}{\lambda \text{ in Å}} \qquad (2)$$

in which E per mole of quanta is given in terms of kilocalories. For example, a system in which an Einstein of 3130A radiation is absorbed has its energy increased by 91 kcal. On a molecular basis it should be noted that absorption of radiation is, at the usual intensities employed, a one quantum process per molecule. Thus, per Einstein of radiation 6.02 x 10^{23} molecules will be photoactivated. As a further example, in acetone the bond dissociation energy of a C—C bond, $D_{C—C}$, is 81 kcal. mole^{-1}; thus a molecule absorbing one quantum of 3130A radiation has received more than enough energy for a homolytic bond rupture.

A photochemical reaction usually involves one or more primary processes. A primary process is generally considered to include the absorption of a quantum of radiation by the reacting molecule and the subsequent reactions of the energy-rich molecule, including dissociation. Certain of these reactions, such as fluorescence, do not lead to chemical change. These are called primary photophysical processes in contradistinction to those leading to permanent chemical change which are designated primary photochemical processes. Important examples of the latter type are illustrated for acetone and acetaldehyde in reactions I and I'.

$$CH_3COCH_3 + h\nu \rightarrow CH_3 + CH_3CO \qquad (I)$$

$$CH_3CHO + h\nu \rightarrow CH_3 + HCO \qquad (I')$$

Absorption of a quantum of radiation, $h\nu$, by the carbonyl chromophore produces a molecule of acetone in an excited state. This energy-rich molecule can rapidly dissociate into free radicals as indicated in (I) or dissipate its energy in photophysical processes such as fluorescence or collisional deactivation. Once formed by (I), the methyl and acetyl radicals undergo a variety of secondary thermal reactions leading ultimately to stable products. Acetaldehyde undergoes an analogous split, (I'), into methyl and formyl radicals but differs from acetone in that it can also undergo a second type of primary photochemical process, (II), which involves an intramolecular rearrangement into the complete molecules methane and carbon monoxide.

$$CH_3CHO + h\nu \rightarrow CH_4 + CO \qquad (II)$$

Here then is one interesting effect of molecular structure on photodecomposition modes. The sole primary photochemical process for acetone is a free radical split, whereas acetaldehyde with an H atom in place of a methyl group undergoes two distinctly different types of primary processes. This matter will be treated in more detail subsequently.

The quantum efficiency of a specific primary photochemical or photophysical process is designated by the symbol ϕ. For example, ϕ_{II} = (Molecules CO formed in (II))/(Total quanta absorbed). Although quantitative estimates of these primary quantum yields are of considerable significance they are difficult to obtain experimentally in a direct, unambiguous manner.

Another symbol often encountered in photochemistry is Φ. In a photochemical process, Φ is the measure of the quantum efficiency of the <u>over-all</u> participation of a specific product or reactant. Thus, in the photolysis of acetaldehyde Φ_{CO} = (Total molecules of CO produced in the <u>entire</u> process)/(Total quanta absorbed).

The primary and the over-all quantum yields for a process differ in this important respect. The sum of the quantum yields for all primary reactions (including fluorescence and deactivation) in a given photochemical process must equal unity (i.e., Σ_ϕ = 1.00), whereas Φ can be as low as 10^{-3} for some processes and as high as 10^6 for others. The essential difference results from the fact that molecules of the product or reactant in question may be involved, not only directly in the primary act, but also in secondary thermal reactions. For example, the aliphatic aldehydes undergo a chain decomposition when photolyzed at elevated temperatures but the aliphatic ketones do not. Thus, Φ_{CO} = 86 for acetaldehyde photolysis at 3130A and 250° whereas, Φ_{CO} = 1.0 for acetone under similar conditions.

A typical sequence of secondary reactions for aldehyde involves methyl radicals formed in primary process (I') reacting with the acetaldehyde substrate to give methane and acetyl radicals.

$$CH_3CHO + CH_3 \rightarrow CH_4 + CH_3CO \qquad (3)$$

$$CH_3CO \rightarrow CH_3 + CO \qquad (4)$$

At temperatures above about 100° the acetyl radicals decompose readily by reaction (4) to produce carbon monoxide and another methyl radical. Thus, the elements of a chain are set up leading to over-all quantum yields of methane and CO that exceed unity. Methyl radical attack on acetone, (5), however,

$$CH_3 + CH_3COCH_3 \rightarrow CH_4 + CH_2COCH_3 \qquad (5)$$

produces the acetonyl radical which is relatively much more stable toward decomposition than acetyl, hence no chain reaction ensues, and Φ_{CO} is unity.

III. EXPERIMENTAL TECHNIQUES

In addition to the experimental problems faced in research on thermal decompositions, a photochemist in the field of reaction mechanisms requires a source of intense monochromatic radiation and a means of quantitatively determining its intensity. He is also faced with acute analytical problems. In quantitative work the amount of photodecomposition is usually of the order of 0.1% to 1%. This prevents appreciable photolysis of the reaction products themselves. Thus with the range of light intensities of steady monochromatic ultraviolet light usually employed (10^{14} to 5×10^{16} quanta/sec.) only micro quantities of products are obtained in runs of several hours duration.

The following brief discussion of some experimental techniques successfully employed in photochemistry touches on these problems and is included here because the material may be of general interest.

Absorption Spectra. The so-called "First Law of Photochemistry" states that in order for light to be effective in producing chemical change it must be absorbed. Thus, the first step in a proposed study of a photochemical process is to obtain the absorption spectrum of the compound being photolyzed. Either vapor-phase or liquid-phase spectra are suitable for determination of the general region in which absorption occurs. The former type shows more structure since solvent effects are absent.

The aliphatic aldehydes and ketones possess in common a relatively weak absorption band in the ultraviolet starting at about 3300-3400A and reaching a maximum in the region of 2775-2875A, depending on the particular compound. Such a curve for acetone is shown in Figure 1. They also show a very strong second absorption band beginning around 2000A and extending into the vacuum ultraviolet or Schumann region. The first band falls in an experimentally accessible region where quartz optics can be readily employed and where mercury arcs give out strong emission lines which are suitable as radiation sources. As a consequence of these fortunate experimental circumstances, the vapor-phase photochemistry of the aliphatic aldehydes and ketones has been widely studied.

Some photochemical studies have been made of α,β-unsaturated aldehydes such as acrolein, a major eye irritant in smog, $CH_2{=}CHCHO$, and ketones. A typical absorption spectrum is included in Figure 1 for trans-methyl propenyl ketone, $CH_3COCH{=}CHCH_3$.

The double bond conjugated to the carbonyl chromophore produces a strong shift to the red (bathochromic), so that the first absorption

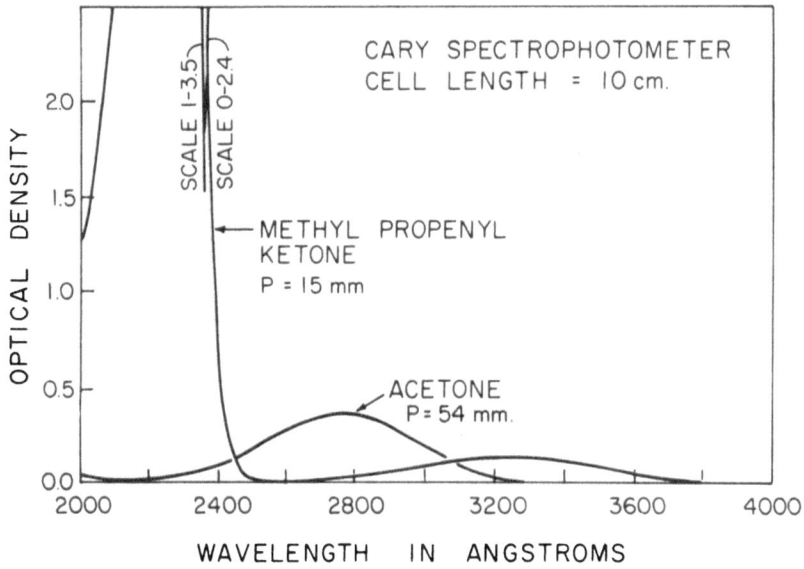

Figure 1. Vapor-phase Absorption Spectra

maximum falls around 3000A and the second about 2200A. Thus, both bands fall in experimentally accessible regions and the photolysis of acrolein, crotonaldehyde and <u>trans</u>-methyl propenyl ketone has been investigated in both absorption regions.

Light Sources. Light sources used in direct photolyses (in contrast to photosensitized reactions) are medium and high pressure mercury arcs enclosed in fused quartz envelopes and operating on either a.c. or d.c. voltage.

Table 1 shows the spectral energy distribution of radiation emitted by a Type A, medium pressure, mercury arc manufactured by the Hanovia Manufacturing Company. This lamp is typical of the type used to obtain line emission spectra. When used in conjunction with the proper filter or monochromator, these lamps are good, stable sources of monochromatic light of adequate intensity for many types of photochemical studies. It is clear from the data that most of the power consumed by the lamp is given off as heat in the infrared or as visible light, but sufficient energy falls in the 3660, 3130, 2652 and 2537A lines to make them particularly useful and widely employed.

High pressure quartz lamps are extremely intense, but in cases where "pure" monochromatic light is desired they suffer from a broadening effect so that the lines tend to be smeared out into a

continuum. Low pressure-high voltage mercury arcs emitting the
2537A mercury resonance line are used in studies of mercury photosensitized reactions.

In order to achieve monochromatic radiation from a multi-line
source such as a mercury arc, it is necessary to use a monochromator
or filter system. A monochromator is convenient to use and is
capable of giving higher purity radiation in the ultraviolet than
chemical filters. Suitable "fast" monochromators give high purity
only at a sacrifice of intensity. These considerations account for
the widespread sue of glass or interference filters in the visible
and ultraviolet, down to and including the 3130A mercury line, and
chemical filters below this.

Actinometry. By definition, in order to determine a quantum
yield, it is necessary at some point to measure the quanta of light
absorbed by the reacting compound, I_a. This = (Total intensity of
radiation falling on the absorbing species) X (Fraction of this
total radiation absorbed). The latter quantity is readily determined
with a simple phototube-galvanometer system. Experimental difficulties arise when I, the total absolute intensity, is to be determined.
The most accurate and direct technique is to determine I by means of
a multijunction thermopile previously calibrated with lamps of standard intensity. Such thermopiles, however, are not readily available and more often chemical actinometers are employed. The latter
technique involves photolyzing first a compound with a known quantum
yield of product. From the known value of Φ product, an analytical
determination of the molecules of product, and a measurement of the
fraction of light absorbed by the reactant, one can evaluate I since

$$I = \frac{\text{Molecules of product}}{\Phi_{product} \text{ X fraction light absorbed by reactant}}$$

Acetone is a convenient actinometer for use in vapor-phase studies
of aldehydes and ketones since Φ_{CO} is 1.00 at temperatures above
120° and wavelengths from 2500A to 3200A. In solution-phase studies,
the photolysis of uranyl oxalate or ferrioxalate solutions have been
the most widely used actinometers, while malachite green leucocyanide
is useful for very low light intensities.

A diagram of a typical reaction system used in vapor-phase
photochemical studies of the type being discussed as shown in
Figure 2. In essence, polychromatic radiation from the water-cooled
quartz-mercury arc A passes into the entrance slits B of the prism
monochromator through the monochromator optics O-P-O, and monochromatic light of the desired wavelength emerges from the exit slit
E. This radiation passes through quartz lens G and a parallel beam
fills the quartz photolysis reaction cell H. Radiation emerging
from H falls on photocell M used for determining the fraction of
incident light absorbed. It is enclosed in water-cooled housing N.

TABLE 1

Radiation Data for the Hanovia Type A, Medium Pressure
Mercury-Quartz Lamp[a]

Region	Wavelength		Intensity (% of Total)
Infrared		Total	55.0
Visible	Red......6234A		0.4
	Yellow...5700A		6.6
	Green....5460A		4.7
	Blue.....4960 and 4358A		4.3
		Total	16.0
Ultraviolet	4045 and 3906A		3.62
	3660A		7.40
	3341A		0.82
	3130A		5.05
	3025A		2.68
	2967A		1.02
	2925,2893 and 2803A		1.00
	2752 abd 2700A		0.65
	2652A		1.78
	2571A		0.68
	2537A		1.95
	2482,2400,2360 and 2300A		1.82
	1942 and 1849A		0.53
		Total	29.00

*Data from Hanovia Research Laboratory

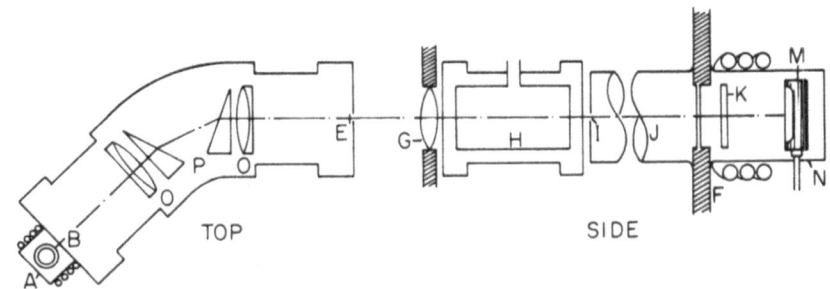

Figure 2.   Diagram of a Type of Optical System Employed for the
Determination of Quantum Yields in Vapor-phase Systems

The reaction cell H is set in an aluminum block furnance I which can be controlled at any temperature from 25° to 400°C., thus enabling one to obtain the effect of temperature on a photochemical process

Analysis of Products. As noted earlier, inquantitative gas phase studies the extent of photodecomposition generally ranges from 0.1% to 1%. This leads to very small amounts of products which form a highly complex mixture. For example, in a typical photolysis of methyl ethyl ketone vapor ($CH_3COC_2H_5$), the total volume of gaseous products pumped off after freezing the system at -79°C might amount to only 30 microliters at S.T.P. (.03 ml of gas). This sample could contain carbon monoxide, methane, ethane, ethylene, propane, butane, and other gases in trace amounts. The residual liquid material at -79°C contains the "condensable products" dissolved in the unreacted substrate. This material is an even more complex and difficult mixture to analyze than the small gas sample. The most effective microanalytical methods for analyses of such mixtures include vapor phase chromatography, mass spectrometry, combined V.P.C.-mass spectrometric techniques and micro ultraviolet, infrared and nmr spectroscopic techniques.

IV. SOME PHOTOCHEMICAL PROCESSES OF SEVERAL SIMPLE CARBONYL COMPOUNDS

Acetone. The simplest ketone probably has been the subject of more gas-phase photochemical investigations than any other compound. Acetone is also a contaminant of urban atmospheres. A simplied version of its photodecomposition in the absence of air includes one free radical primary process and a series of secondary thermal reactions of the radicals produced therein.

The Primary Process:

$$CH_3COCH_3 \ + \ h\nu \rightarrow CH_3 \ + \ CH_3CO \tag{I}$$

Some Typical Secondary Reactions:

A. Decomposition

$$CH_3CO \rightarrow CH_3 \ + \ CO \tag{4}$$

B. Abstraction

$$CH_3 \ + \ CH_3COCH_3 \rightarrow CH_4 \ + \ CH_2COCH_3 \tag{5}$$

C. Combination

$$2CH_3 \rightarrow C_2H_6 \tag{6}$$

D. Disproportionation

$$CH_3 + CH_3CO \rightarrow CH_4 + CH_2CO \tag{7}$$

Primary process (I) is efficient and reaction (4), the dissociation of the acetyl radical, occurs rapidly at temperatures above 100°C. Hence, at elevated temperatures the primary process is effectively I_B.

$$CH_3COCH_3 \rightarrow 2CH_3 + CO \tag{I_B}$$

Acetone photolysis at elevated temperatures thus furnishes a clean source of methyl radicals, and this technique has been widely used.

It is interesting to note that no intramolecular primary photochemical process giving CO has been demonstrated for acetone, despite an extensive search in which iodine inhibition and deuterium labeling techniques have been employed.

Some typical secondary reactions of importance, reactions 4-7, are given to illustrate the variety of products obtained from such a "simple" system. Many other products have also been found which adds to the complexity of the situation.

Aldehydes. These are a particularly important constituent of automobile exhaust. Thus, while control devices are reducing hydrocarbon emissions, the amounts of aldehydes released to the atmosphere generally are increasing! This point is made here because (a) it is often overlooked and (b) because it illustrates the importance of understanding the photochemistry of this class of compounds. The latter has been reviewed in detail by Calvert and Pitts[1] and recent papers by Altshuller and co-workers[11], and will not be discussed here.

Among the most important of the aldehydes in atmospheric chemistry are formaldehyde, H_2CO; acetaldehyde, CH_3CHO; and benzaldehyde, C_6H_5CHO. For example, photolysis of formaldehyde leads in part to the formation of H atoms and formyl radicals, $H_2CO + h\nu \rightarrow H + HCO$, while an important primary process in acetaldehyde photolysis produces methyl and formyl radicals

$$CH_3CHO + h\nu \rightarrow CH_3 + HCO \tag{I'}$$

Species such as H, CH_3, CHO, etc. are highly reactive and can initiate, or participate, in highly complex reaction sequences leading to a variety of products. This is particularly true in urban atmospheres in which NO and NO_2 are present. The important

pollutant PAN, $CH_3C\overset{\displaystyle\nearrow O}{\underset{\displaystyle\searrow OONO_2}{}}$ may be formed by one of these mechanisms.

V. SOME ASPECTS OF THE PHOTOCHEMISTRY OF NITROGEN DIOXIDE

Nitrogen dioxide, NO_2, is, structurally speaking, a simple non-linear triatomic molecule. However, from both an "academic" and an "applied" point of view, its chemical and physical properties are complex and unusually interesting. It is a free radical (the dimerization to N_2O_4, nitrogen tetroxide, is a classic case of chemical equilibrium), deeply colored, very reactive, toxic, and apparently the chief chemical agent to photochemically open Pandora's Box and generate the witch's brew of matter characteristic of photochemical air pollution of the type found in the Los Angeles Basin. Thus, in class studies, Haagen-Smit and his collaborators first pointed out about two decades ago that oxidation processes in the oxides of nitrogen-hydrocarbons-sunlight system produced the manifestations of photochemical air pollution (except for some aspects of aerosol formation which, though important, will not be dealt with here.) Furthermore, it is significant that Haagen-Smit and Fox stated as far back as 1956, "The present control measures are directed mainly toward reduction of hydrocarbon, while nitrogen oxides from high temperature combustion sources continue to increase. This increase in nitrogen oxides demands greater and greater efficiency in hydrocarbon recovery. Since the reduction of hydrocarbon emission will be only partial, it is essential to study the other component of the smog-forming system, nitrogen oxides, and to institute engineering research necessary for the drastic reduction of nitrogen oxides released by combustion processes."[10]

The interrelationships between the various oxides of nitrogen such as NO, NO_2, N_2O, NO_3, N_2O_4, etc., etc., symbolized by NO_x, are complex, particularly in the presence of light, oxygen, ozone, and organic matter. However, it is now clear that of these oxides only NO_2 is both present in significant amounts and also a strong absorber of sunlight, the energy source required for initiation of photochemical air pollution (see Figures 3 and 4). Furthermore, it photodissociates to give highly reactive fragments (oxygen atoms) capable of initiating the thermal reactions responsible for the overt physiological manifestations of smog such as eye irritation and plant damage.

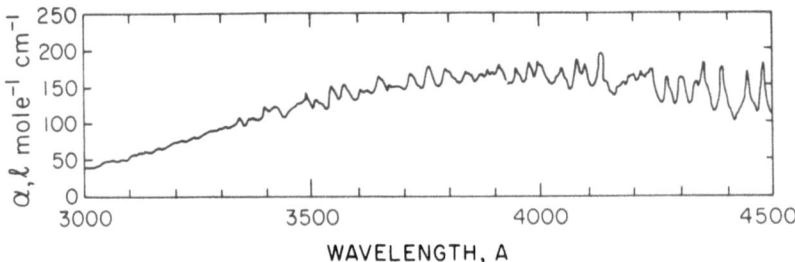

Figure 3. Absorption Coefficients of Nitrogen Dioxide in the Region 3000-4500A.[1]

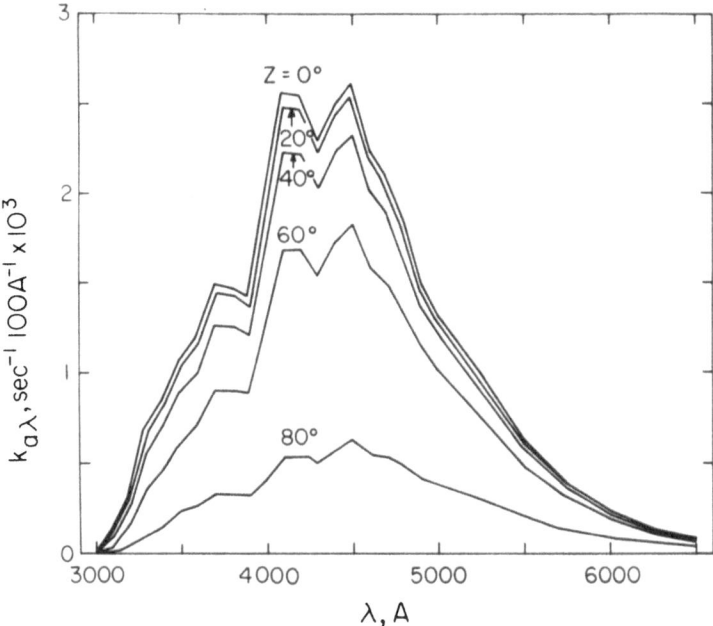

Figure 4. Absorption of Solar Radiation by Nitrogen Dioxide.[2]

While the importance of nitrogen dioxide in air pollution was recognized years ago, until recently basic information concerning details of its photochemistry in both the mm. Hg pressure range commonly employed by most photochemists and kineticists and the ppm. and pphm. concentration ranges encountered in polluted atmospheres was relatively sparse. This has recently been reviewed by Schuck and Stephens [9] as well as in Leighton's monograph [2] and the papers and reviews by Pitts et al.[4,7], Altshuller and Bufalini[5] and Haagen-Smit and Wayne [8] and will not be considered in detail here. We shall simply discuss some aspects of the photochemistry of NO_2-O_2 mixtures as they further illustrate principles of photochemistry of direct application to photochemical air pollution.

Photochemistry of NO_2 in the Visible Region ($\lambda > 4358A$). The vapor phase photochemistry of nitrogen dioxide in the visible portion of the solar spectrum, i.e. at 4358A and longer wavelengths, is a good example of the mechanism consisting only of photophysical processes. In this region irradiation of either pure NO_2 or of NO_2 in air with hydrocarbons added, does not lead to permanent chemical changes in the system. This is particularly pertinent to air pollution since a large fraction of the sunlight absorbed by NO_2 falls in the region $\lambda \geq 4358A$ (cf. Figure 4). In this region the mechanism may be written as,

Mechanism A ($\lambda \geq 4358A$)

Absorption:

$$NO_2 + h\nu \rightarrow NO_2^* \tag{III}$$

Collisional Deactivation:

$$NO_2^* + NO_2(\text{or M}) \rightarrow NO_2 + NO_2(\text{or M}) \tag{8}$$

Fluorescence:

$$NO_2^* \rightarrow NO_2 + h\nu' \tag{9}$$

Resonance Transfer:
(Intersystem Crossing)

$$NO_2^* \rightarrow NO_2^{**} \tag{10}$$

Collisional Deactivation:

$$NO_2^{**} + NO_2(\text{or M}) \rightarrow NO_2 + NO_2(\text{or M}) \tag{11}$$

Fluorescence:

$$NO_2^{**} \rightarrow NO_2 + n\nu'' \tag{12}$$

Step 10 is a _resonance_ _transfer_ (or _intersystem_ _crossing_) from a higher excited electronic state to a lower one. It is a general type of process highly important in photoactivated systems. The species M in Equations 8 and 11 is a _third_ _body_ which may take up the electronic excitation energy of NO_2^* or NO_2^{**} and convert it to thermal energy without chemical change.

Discussion of these spectroscopic processes and the details of the electronic structure of NO_2 are beyond the scope of this paper. The essential feature of the overall Mechanism A is, however, that at $\lambda \geq 4358A$, nitrogen dioxide appears to be photochemically inert, either alone or in the atmosphere.

Photochemistry of NO_2 in the Ultraviolet. For the photolysis of pure NO_2 at mm. pressures and 3130A the _overall_ photochemical process is given by

$$2 NO_2 + h\nu \rightarrow 2 NO + O_2 \tag{13}$$

and $\Phi_{O_2} = 1.0$, $\Phi_{NO_2} = 2.0$.

Equation (13) does not, however, specify the detailed mechanism by which oxygen is formed. Thus, one can postulate the following two mechanisms, differing greatly in their chemical implications, but both resulting in the overall reaction, Equation (13).

Mechanism B (3130A)

Primary Process:

$$NO_2 + h\nu \rightarrow NO + O \, (^3P) \quad\quad\quad (IV)$$

Secondary Reaction:

$$O + NO_2 \rightarrow NO + O_2 \quad\quad\quad (14)$$

Overall Reaction:

$$2 \, NO_2 + h\nu \rightarrow 2 \, NO + O_2 \quad\quad\quad (13)$$

Mechanism C

Primary Process:

$$NO_2 + h\nu \rightarrow NO_2^* \qu\quad\quad\quad (III)$$

Secondary Process:

$$NO_2^* + NO_2 \rightarrow 2 \, NO + O_2 \quad\quad\quad (15)$$

Overall Process:

$$2 \, NO_2 + h\nu \rightarrow 2 \, NO + O_2 \quad\quad\quad (13)$$

Mechanism B involves the direct photodissociation of NO_2 into nitric oxide and highly reactive oxygen atoms, both in their ground electronic states, ^3P for the oxygen atoms. Thus, in pure NO_2, oxygen atoms formed in the primary photochemical process IV react by the secondary, thermal process, equation (14), with the excess NO_2 present to give oxygen and another molecule of nitric oxide. However, with only trace amounts of NO_2 in the atmosphere present with a huge excess of oxygen, the thermal reaction to form ozone, equation (16), predominates over reaction (14).

$$O + O_2 + M \rightarrow O_3 + M \quad\quad\quad (16)$$

The species M is a "third body" which must be present to siphon off the energy formed in the highly exothermic process (16) otherwise the hot O_3 molecule would immediately dissociate back onto $O + O_2$.

In the atmosphere M is usually N_2 or O_2 simply because they are the major constituents. If the oxygen atom collides with another reactive species in the atmosphere before ozone formation occurs, an olefin, for example, then it can add to the double bond and directly lead to the formation of epoxides, aldehydes, etc. However, in polluted atmospheres the hydrocarbons are present in trace amounts relative to oxygen, and ozone formation is the chief fate of oxygen atoms formed from NO_2 photolysis at 3130A. The ozone formed in (16) is relatively stable to thermal decomposition. It can exist long enough to react with hydrocarbons or other secondary contaminants to form or be a precursor to many of the host of compounds that characterize typical photochemical "smog". It can also absorb radiation and photodissociate[6].

Mechanism C is an "excited molecule" type in which the photoactivated nitrogen dioxide, symbolized as NO$\overset{*}{2}$, exists long enough in a reactive state (without losing its electronic energy by fluorescence or collisional deactivation) to react with a nonactivated molecule of NO_2 to give molecular oxygen directly. There is no intermediate formation of oxygen atoms as in Type B. Thus, secondary reactions initiated by the Type C Mechanism would be expected to be much less probably and if they occurred, different in character and products than in Mechanism B.

In fact, there is now good experimental and spectroscopic evidence that the photodissociative Mechanism B is the correct one at 3130A. Thus at 3130A the absorption spectrum exhibits a diffuse continuum indicating dissociation, no fluorescence is observed, and several tracer experiments with added O_2 show isotope scrambling[4]. Such scrambling would not occur if the nonphotodissociative mechanism were the correct one.

Since the overall quantum yield of oxygen formation, Φ_{O_2} is 1.0 at 3130A, Mechanism B implies that the primary yield of oxygen atoms from primary process IV is unity at this wavelength and drops to zero at 4358A. This effect is shown in Figure 5 which shows both the quantum yield of molecular O_2 and the quantum yield of isotope scrambling as a function of wavelength of exciting light. The two curves are remarkable similar indicating that the photodissociative Mechanism B is the sole mechanism of photodecomposition of NO_2 not only at 3130A but over the entire range 4358-3130A.

It is of interest to note, in conclusion, that in the far ultraviolet, NO_2 seems to dissociate into nitric oxide and an oxygen atom in its first _excited_ state, $O(^1D)$. It is important to note that there are significant differences in reactivity of oxygen atoms in the excited singlet (^1D) versus ground state triplet, (^3P). This point is significant not only as academic research but also because it has important implications for high altitude and outer space studies where high fluxes of solar radiation in the vacuum ultraviolet (2000

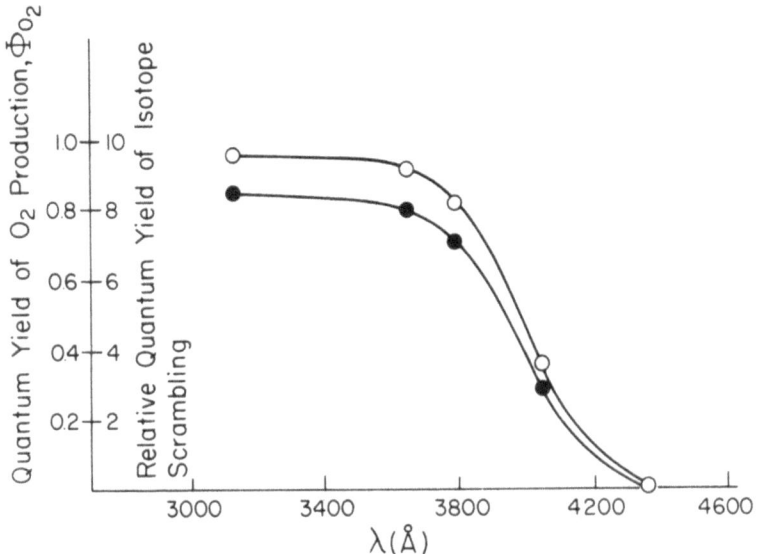

Figure 5. A Comparison of the Overall Quantum Yields of Oxygen
    Production, Open Circles, and the Extent of Dissociation into
    NO and O Atoms (i.e. the Relative Quantum Yields of Isotope
    Scrambling), shaded circles, as a function of the wavelength
    of the incident light (4).

to about 1200A and lower) are encountered.

REFERENCES

A. Fundamental

1. J. G. Calvert and J. N. Pitts, Jr., Photochemistry, John
Wiley and Sons, Inc., New York, 1966, 899 pages.
2. P. A. Leighton, Photochemistry of Air Pollution, Academic
Press, New York, 1961, 300 pages.
3. J. N. Pitts, Jr., "Relations Between Molecular Structure
and Photodecomposition Modes", J. Chem. Ed., 34, 112 (1957).
4. James N. Pitts, Jr., James H. Sharp and Sunney I. Chan,
"Effects of Wavelength and Temperature on the Primary Processes of
Nitrogen Dioxide and a Spectroscopic-Photochemical Determination of
the Dissociation Energy", J. Chem. Phys., 40, 3655 (1964).

B. Applications to Photochemistry

    5. A. P. Altshuller and J. J. Bufalini, "Photochemical Aspects
of Air Pollution: A Review", Photochem. Photobiol., 4, 97 (1965).
    6. J. N. Pitts, Jr., A. U. Khan, E. Brian Smith, and R. P.
Wayne, "Singlet Oxygen in the Environmental Sciences. II. Singlet
Molecular Oxygen and Photochemical Air Pollution", Environ. Sci.
Technol., 3, 241 (1969).
    7. J. N. Pitts, Jr., "Environmental Appraisal: Oxidants,
Hydrocarbons and Oxides of Nitrogen", Air Pollution Control Associa-
tion Symposium Proceedings, June 22-26, 1969, To appear in a special
issue of the Journal of the Air Pollution Control Association.
    8. A. J. Haagen-Smit and L. G. Wayne, "Atmospheric Reactions
and Scavenging Processes" in Air Pollution, Vol. 1, A. C. Stern,
Ed., Academic Press, New York, 1968.
    9. E. A. Schuck and E. R. Stephens, "Oxides of Nitrogen" in
Advances in Environmental Sciences, Vol. 1, J. N. Pitts, Jr. and
R. L. Metcalf, Eds., Interscience, New York, 1969, p. 73.
    10. A. J. Haagen-Smit and M. M. Fox, "Ozone Formation in Photo-
chemical Oxidation of Organic Substances", Ind. Eng. Chem., 48, 1484
(1956).
    11. A. P. Altshuller, I. R. Cohen and T. C. Purcell, "Photo-
oxidation of Propionaldehyde at Low Partial Pressures of Aldehyde",
Science, 156, 937 (1967).

DETECTION AND ANALYSIS OF ATMOSPHERIC POLLUTANTS

Peter K. Mueller, Chief

Air and Industrial Hygiene Laboratory

State Department of Public Health, Berkeley

I. INTRODUCTION

There are several objectives for air quality measurements and each presents a unique set of requirements to be considered in the selection of sampling sites, sampling frequencies and methods of analysis. Firm concepts about rational approaches are still being developed (76, Morgan, Saltsmann, Percy and Hildebrandt, Bell, McGuire and Noll, and Whitby, Proceedings 11th Methods Conference*), but these authors have already presented useful guidelines.

This chapter is intended to be an annotated outline of the many complex aspects involved in the reliable assessment of air quality. This chapter should help to guide the planning of air monitoring projects. It presents sufficient detail about several frequently-used procedures to create an awareness for needs of equipment, manpower and new capability. Examples and alternatives are presented to provide perspectives. The current literature is amply referenced for those wishing to delve into the details of specific topics.

II. ATMOSPHERIC COMPOSITION

The atmosphere is a complex mixture of gases and particles. The disperse phase which is in a molecular or submolecular state is the gas. An aerosol is a suspension of molecular aggregates (solid or liquid particles) in gas.

Many gases and particles in air derive directly and uniquely from human activities. Cases are O_3, H_2O, CO_2, CO; CH_4, paraffinic,

*References 77 through 82.

olefinic and aromatic hydrocarbons; organic carbonyl compounds,
organic acids, heterocyclic nitrogen, and sulfur compounds; organic
sulfur, halogen and phosphorus compounds; NH_3, NO, NO_2; H_2S, SO_2.
Constituents of man-made particles include: minerals such as carbon,
asbestos, talc, silicates, sulfates, nitrates, chlorides, and
fluorides compounded with Fe, Pb, Zn, Cu, Mn, Ni, As, V, Be, Sr,
Ba, Cd, Co, Hg, Se in addition to Na, K, Mg, Ca and Al; liquids
such as H_2O with dissolved substances; H_2SO_4, HNO_3, oils; organics
of the various classes adsorbed on or absorbed in minerals and
liquids, and compounds and aggregates generated by combustion or
photochemically from gas phase compounds.

Among the above-named substances there are many not necessarily anthropogenic, which creates the need to separate "background" from controllable levels of pollution. The gases include:
N_2, O_2, O_3, noble gases; H_2O, CO_2; CH_4, hydrocarbons such as
terpenes, alcohols, carbonyl compounds, and organic acids; NH_3,
N_2O, NO, NO_2; H_2S, SO_2. The nonanthropogenic constituents of
particles in air include: minerals such as silicates, carbonates,
chlorides and sulfates generally compounded with Na, K, Mg, Ca,
Fe or Al. Organics such as pollen, spores, microorganisms, seeds,
and photochemically generated polymers.

The concentration of atmospheric constituents other than O_2
and N_2 are given in Table 1. Figures 1 and 2 are charts for
converting μ g/m^3 to ppm. The concentrations of many substances
commonly found in urban air are compared in Table 7-2. Table 7-2a
gives what may be a typical composition of particulate matter in
Los Angeles. Much remains to be found out about specific components and their origins.

III. SAMPLE GATHERING AND STABILITY

A. Sample-gathering techniques consist of those in which
passive samplers are used (called in-situ collection) and those
involving active samplers. In-situ methods include the following:

1. Sieves may be used in the collection of particles having
a cross-sectional diameter greater than 5 microns.

2. Impactors may be used for particles having an equivalent
diameter greater than 5 microns. The Rotorod (8) is an example
of this type of sampler.

3. Various absorption surfaces are used in the collection
of reactive gases. Among these are:

a. PbO_2 candles for total sulfation (9, 48, 69*, 70*, 71*)

*Refer to promising new approaches

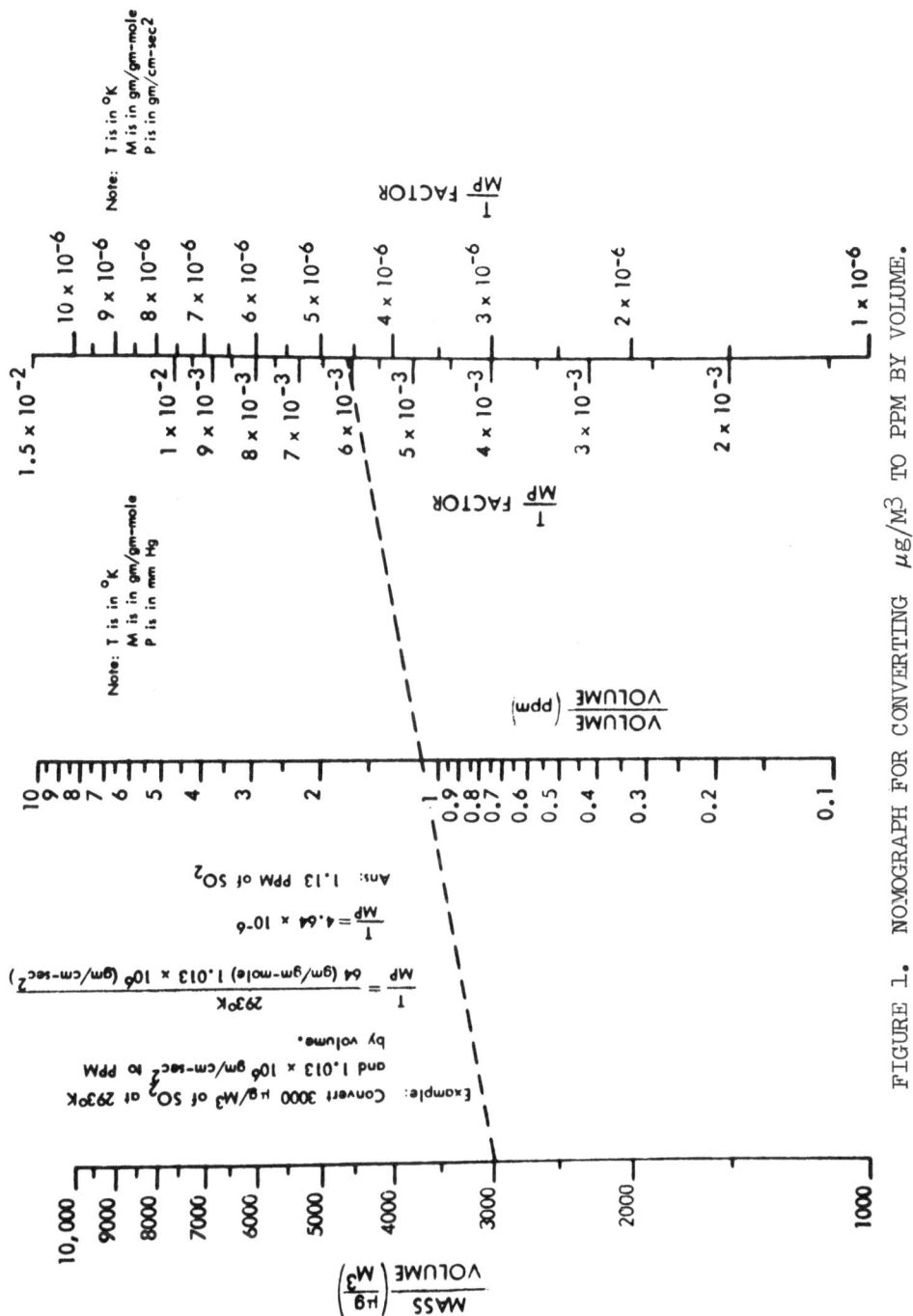

FIGURE 1. NOMOGRAPH FOR CONVERTING $\mu g/M^3$ TO PPM BY VOLUME.

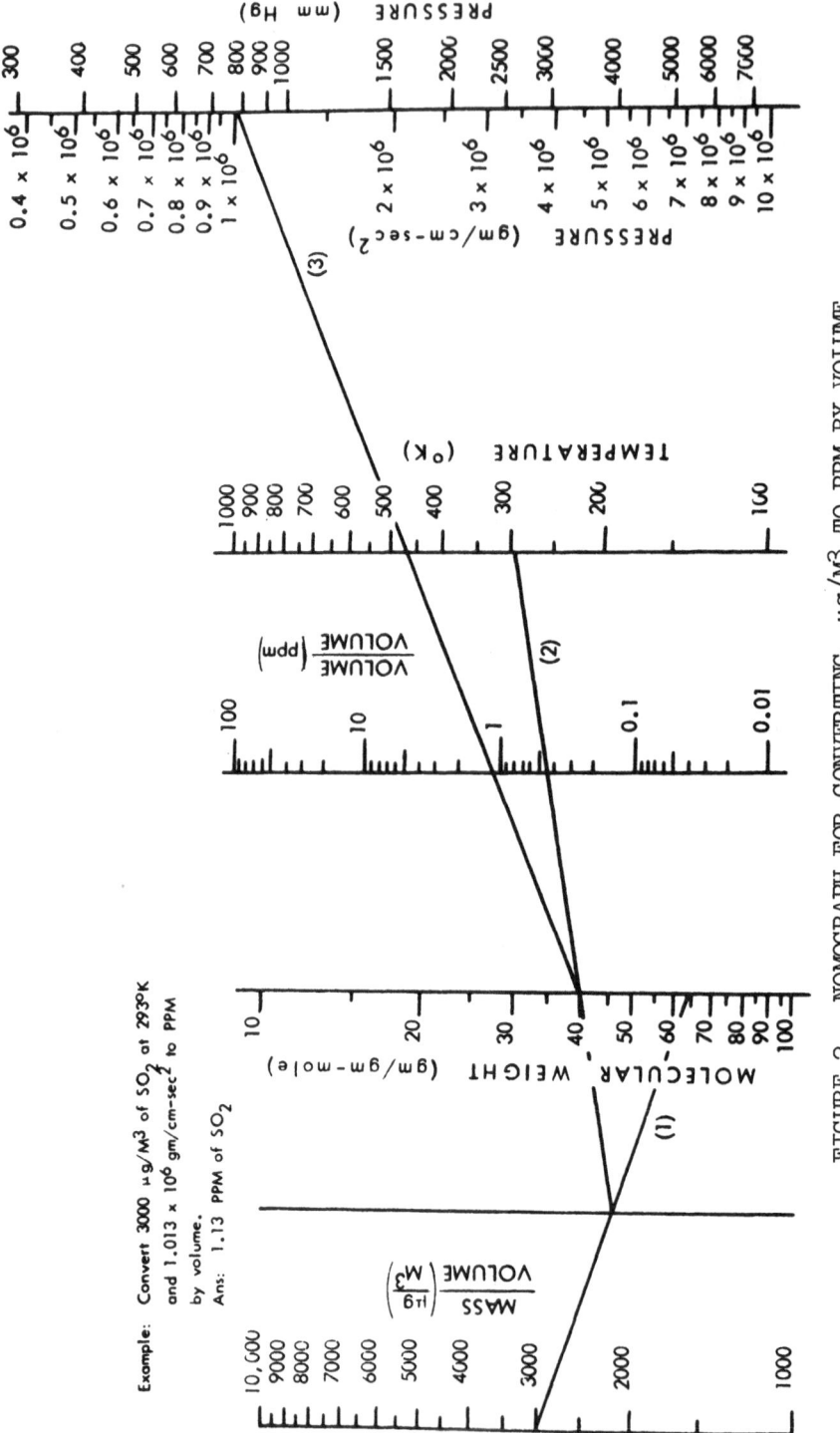

FIGURE 2.   NOMOGRAPH FOR CONVERTING $\mu g/M^3$ TO PPM BY VOLUME.

  b. Lime-impregnated paper for fluorides (10)

  c. Pb acetate impregnated tiles for sulfides (11)

  d. Rubber cracking test for O_3 (12)

  e. Material surfaces (13, 22)

 4. Certain pollutants can be estimated by long-path observation of changes in the following types of radiation:

  a. Visibility (14, 15)

  b. Infrared radiation (16, 17, 21)

  c. Infrared or ultraviolet absorption (Table 3)

  d. Microwave absorption

 5. The occurrence of malodors can be based on reports from
citizens (49).

B. Aerosol Gathering Techniques

 1. Inertial techniques are used to separate particles greater
than $0.5\,\mu$ equivalent diameter by inducing a change in the velocity
of the aerosol. The particles may be deposited on solid surfaces
or suspended in liquids.

 2. Gravitational techniques are used to deposit particles
greater than 1 micron on a surface by sedimentation. Particles
down to 0.05 microns can be separated from an aerosol by the
induced gravity of a centrifuge.

 3. The gradient techniques induce a directed motion of
particles. These techniques include:

  a. Electrophoresis (7) 0.001 to 5 μ equivalent diameter

  b. Thermophoresis < 0.001 to 10 μ equivalent diameter

  c. Diffusophoresis (vapor concentration gradient)

  d. Photophoresis

 4. Filtration methods, by which particles are removed by
sieving, diffusion, inertia and electrostatic charge, are
applicable to particles from 0.001 to 100 μ , depending on face
velocities, flow rates, filter medium, pressure drop and aerosol

concentration. Many filters display a characteristic size of
maximum particle penetration with higher efficiency for both larger
and smaller particles. A recommended methodology for testing
filter performance has been published (37). Some performance
characteristics of air sampling filters have been published (3,51).

TABLE 1. ATMOSPHERIC GASES

Gas		Atmospheric Values		
Name	Formula	Ground Level		Residence Time
		ppm	$\mu g/m3$ STP	
Argon	Ar	9300	1.6×10^7	--
Neon	Ne	18	1.6×10^4	--
Helium	He	5.2	920	2×10^6 yrs.
Krypton	Kr	1.1	4100	--
Xenon	Xe	0.086	500	--
Water vapor	H_2O	$(0.4-400) \times 10^2$	$(3-3000) \times 10^4$	10 days
Ozone	O_3	$(0-5) \times 10^{-2}$	0-100	2 days
Hydrogen	H_2	0.4-1.0	36-90	--
Carbon dioxide	CO_2	$(2-4) \times 10^2$	$(4-8) + 10^5$	4 years
Carbon monoxide	CO	$(1-20) \times 10^{-2}$	$(1-20) \times 10^1$	0.3 yrs.
Methane	CH_4	1.2-1.5	$(8.5-11) \times 10^2$	100 yrs.
Formaldehyde	CH_2O	$(0-1) \times 10^{-2}$	0-16	--
Nitroud oxide	N_2O	$(2.5-6.0) \times 10^{-1}$	$(5-12) \times 10^2$	4 years
Nitrogen dioxide	NO_2	$(0-3) \times 10^{-3}$	0-6	--
Ammonia	NH_3	$(0-2) \times 10^{-2}$	0-15	--
Sulfur dioxide	SO_2	$(0-20) \times 10^{-3}$	0-50	5 days
Hydrogen sulfide	H_2S	$(2-20) \times 10^{-3}$	0-30	40 days
Chlorine[a]	Cl_2	$(3-15) \times 10^{-4}$	1-5	--
Iodine[b]	I_2	$(0.4-4) \times 10^{-5}$	0.05-0.5	--

[a]Gaseous Cl compound; not proven to be Cl_2
[b]Fraction of I_2 likely to be adsorbed on aerosols

C. Gas Sampling

1. Storage sampling means moving a potion of ambient air into
a container and closing it for transportation to the locale of

analysis. There is a danger of taking the sample so rapidly that
it is not representative of the atmosphere. There are different
techniques for doing this:

a. Compressed storage is used for taking large volumes
by pressurizing the sample in a gas cylinder. Compression
frequently creates chemical changes in the sample and its use
appears to be limited to studies of isotope abundances.

b. Rigid containers of glass or stainless steel are
generally inert. Sampling is done either by initially evacuating
and filling or by flushing. Absolute minimization of sample-
surface interactions are obtained by repeated evacuation and
filling prior to a final filling. It is advisable on reuse to
reserve containers for specific substances.

c. Non-rigid containers are plactic bags with many
different types of valved closures. The choice of material depends strongly on pollutants to be collected, their concentration
and the time between the sampling and the analysis. Changes in
the sample may result from permeation through the walls, out gas-
sing from the plactic, reactions, adsorption and condensation on
the walls. The available information on plastic bag performance
has been summarized (24, 75). Bags made from Type 18 Saran Film
(25), Aluminized Scotchpak (26) and 2 mil Teflon or Tedlar (27)
are among the most frequently used.

Samples of approximate volume sufficient for analytical requirements may be taken once every few minutes or
accumulated intermittently. Aliquots of precisely known volume
are then used for analysis. Alternatively, the atmosphere may be
gathered at a constant rate for a given period by a technique
illustrated in Figure 3. This results in a sample volume
accurately integrated with respect to time.

d. Variable volume containers are special gas
syringes which are useful for direct transfer to gas chromatographs
or for the addition of measured volumes of gas and liquid phase
reagents. Thus, their use minimizes transfer errors. The fragility
of syringes generally had made them popular in use only where time
and travel between sampling and analysis are short. The convenient
sampling volume is > 100 ml. This is sufficient for the analysis of
atmospheric hydrocarbons. Samples larger by a factor of 10 to
10^2 are needed for the chemical analysis of polar substances.

e. Continuous sampling is associated with continuous
atmospheric analysis instruments. In such systems air, at a
precisely regulated constant rate, is brought into contact, with
or without suitable treatment, with a concentration transducer.

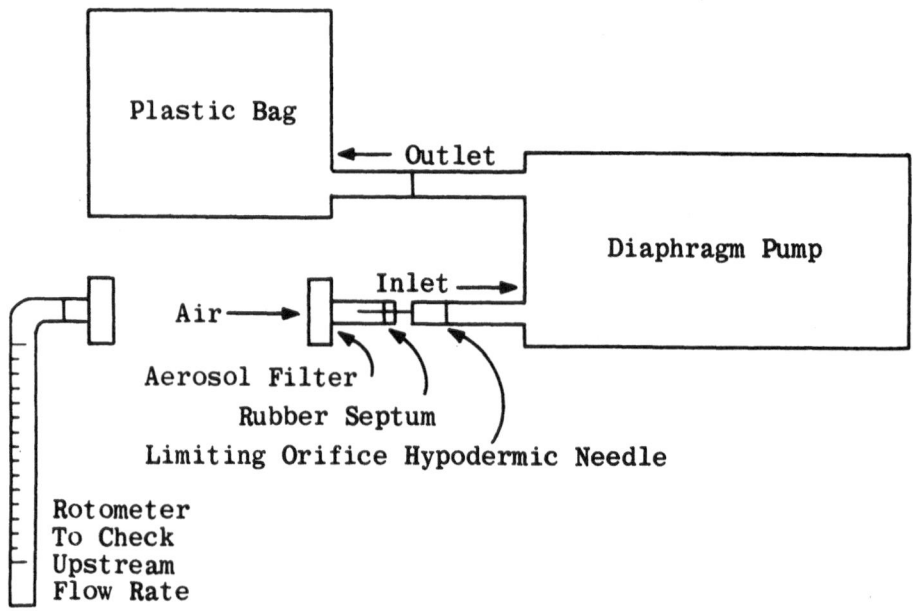

FIGURE 3.   ASSEMBLY FOR INTEGRATED HYDROCARBON SAMPLING.

Required treatment frequently involves absorption in a reagent and subsequently the transducer is made to detect the reaction products.

2. Absorption is the process by which a component in the gas phase is transferred into the bulk of a liquid or solid phase. This process is usually accompanied by a chemical reaction which fixes the component in a new state making the transfer essentially irreversible. At a constant absorption efficiency, the amount absorbed is proportional to the gas sample volume.

The absorption processes used analytically are not always irreversible. Care must be exercised to avoid an erroneous assumption. When the process is reversible a steady state must be established, and then the amount absorbed is independent of sample volume and dependent only on the concentration of the gas phase component. Reversibility is a function of absorbent selection, gas phase concentrations, gas phase composition and scrubber design (bubbler, concurrent or countercurrent principles).

To avoid assumptions concerning absorption in specific sampling situations, absorber calibrations and performance testing should include direct measurement of entering and exiting concentrations, amounts retained by tandem absorbers and comparison to an independent complete absorption method. Generalized and specific mathematical discussions have been published (28, 29, 30, 51, 52, 53, 54).

The efficiency of gas to absorber transfer is, in addition, a function of the stability of the gas in the system prior to absorption, solubility, sample flow rate to absorbant volume and absorbant surface to gas volume ratios. At high solubilities the last factor becomes negligible. Performance of different types of scrubbers have been discussed by Hendrickson (50). Apparatus for sampling SO_2, O_3, and NO_2 illustrates some of the practical variables (Fig. 4). The relative solubilities in water of commonly occurring gases are given in Table 4.

    a. Liquid absorbers

Contact between gas and liquid is achieved by ventilation, spraying or bubbling. Whenever possible, designs are tailored for specific situations to optimize efficiency with response time, diminution of interferences space and cost.

Bubblers are the most common type of contactors. Maximum ratio of sample flow rate to absorbent volume consistent with reasonably complete absorption is an essential requirement. For foaming reagents adequate free space is needed. Dispersers are made from sintered glass. Little is to be gained in effeciency from generating smaller bubbles with very fine frits. Bubble

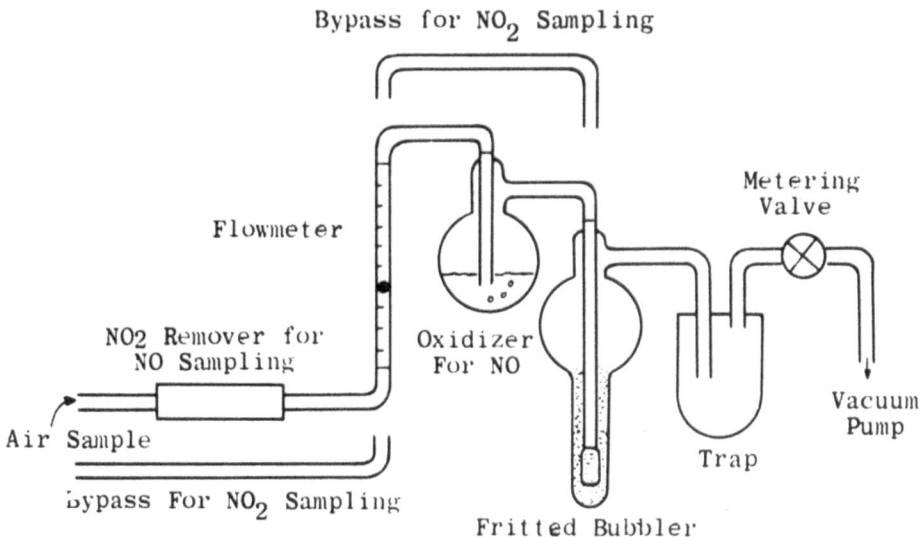

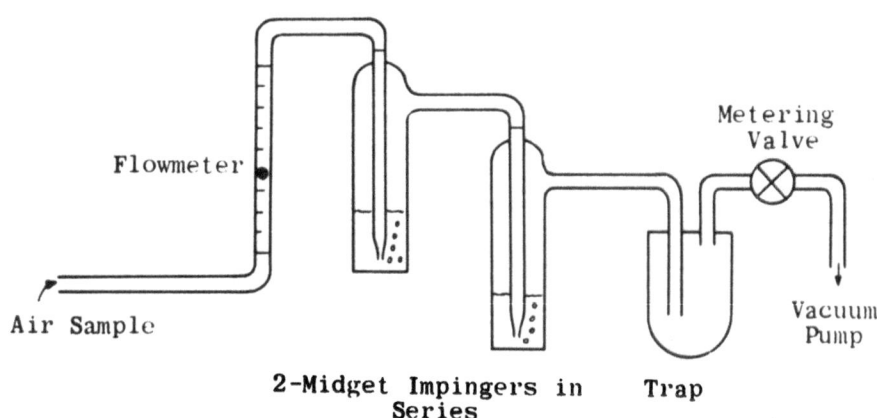

FIGURE 4.   APPARATUS FOR SAMPLING NO AND NO_2, O_3, AND SO_2.

coalescence tends to negate any gain in bubble surface from an
initially smaller size and by the increase in pressure drop
through the finer frit. More is gained by increasing contact
time through increasing the height of the liquid above the frit.

Frits cannot be used with O_3, which is catalytically de-
composed on the surfaces. For this purpose open ended tube
bubblers are used. For highly soluble substances, i.e. SO_2 is
acidified H_2O_2, the liquid surface may be kept below the gas inlet
tube. Interferences from less soluble substances are thereby
minimized.

Continuous air analyzing instruments utilize concurrent
or countercurrent air and liquid flows in thin tubes with or
without packing. When the absorbing solution reacts to form a
colored product for photometry 100% absorption can be established.
These can subsequently be adjusted to provide direct concentration
read-out on absorbance chart paper (32).

b. Solid absorbers

Those passively used have been covered (Section III A 3).
For active sampling (moving known volumes) packed beds or impregnated filters are used.

There are many examples of packed beds including packings
of solid relatively inert substrates coated with liquid layers.
Alkyl lead compounds (e.g. tetraethyl lead)have been sampled in
air using a bed of iodine crystals preceded by a filter to remove
particles (55).

As more is learned about the performance of impregnated
filters their use will become increasingly popular. Membrane
filters impregnated with $KHCO_3$ or KOH in triethanolamine have
been used for sampling SO_2 down to 10 ppb. Membrane filters impregnated with lead acetate have been used for sampling H_2S down
to 10^{-3} ppm. Cellulose filters impregnated with Na_2CO_3 are used
to sample acid gases and mists. Efficiencies depend on filter
media, absorbent formulation, flow rate, temperature, relative
humidities and substances in air besides the gas to be sampled.
Impregnated filters are also being used as scrubbers to remove
interferences. Adams (56) has selectively separated several
sulfur compounds in gases by use of a series of different filters
(Table 5).

3. Condensation

Low temperature condensation methods utilize liquid hydrogen
to totally condense air (rare), liquid nitrogen (undesirable

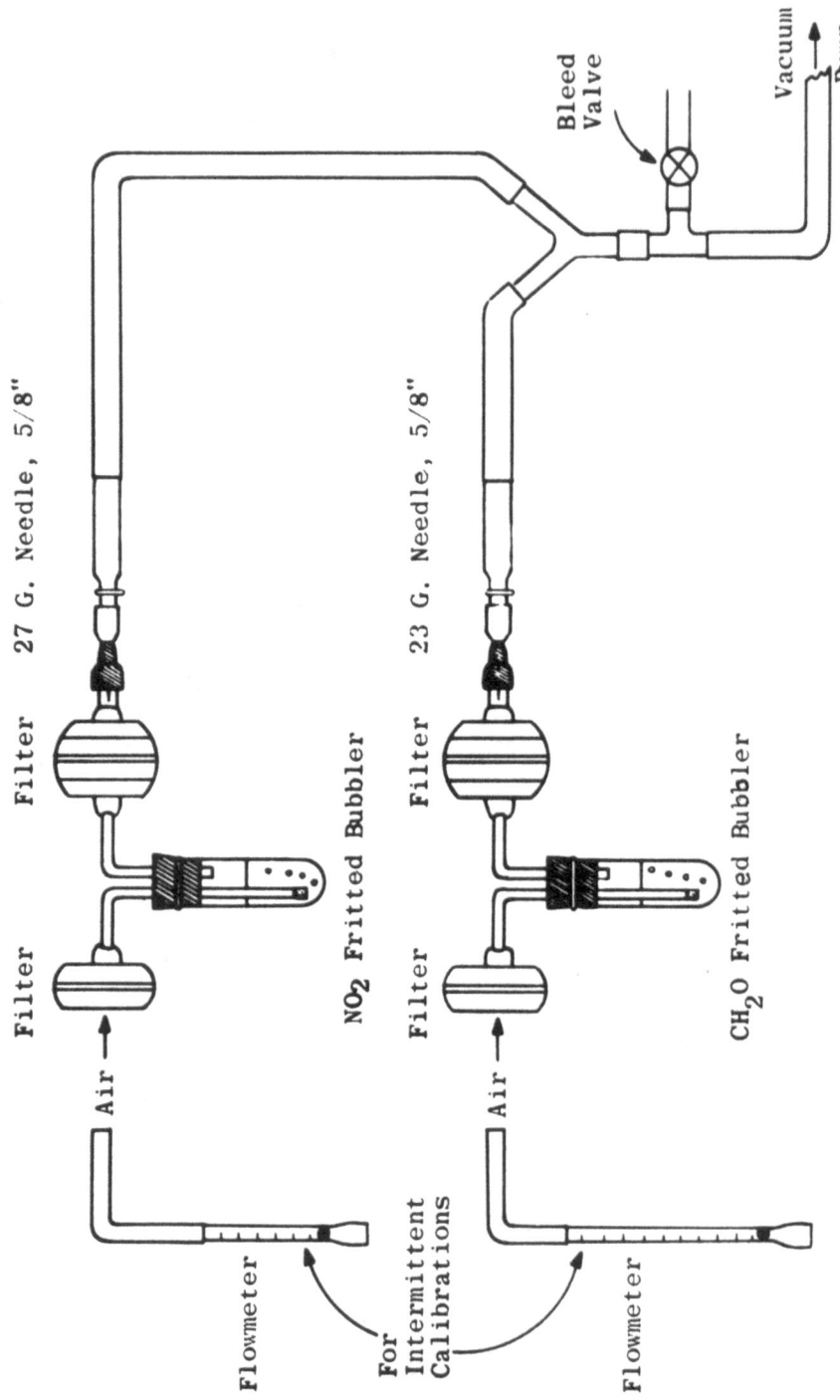

FIGURE 5. TYPICAL NO$_2$ AND CH$_2$O SAMPLING APPARATUS.

TABLE 2. APPROXIMATE AVERAGE CONCENTRATIONS
IN URBAN AIR[2]

Substance	$\mu g/1000m^3$	Substance	$\mu g/1000m^3$
CO_2	5×10^8	Oxidants at O_3	2×10^4
CO	8×10^6	Particulate acid as H_2SO_4	1.4×10^4
CH_4	3×10^6	Acrolein	1.4×10^4
C_2H_2	1×10^5		
C_6H_6	1×10^5	C_6H_6 soluble particulates	1×10^4
Suspended particulates	1×10^5	H_2S	5×10^3
SO_2	8×10^4	Large aliphatic HC	3×10^3
HCHO	7×10^4	Large n-alkanes $(C_{14}-C_{30})$	5×10^2
NO_2	6×10^4	n-Tricosane	8×10^1
NO	4×10^4	Benzo(a)pyrene	6
Phenols	2×10^4	Benz(c)acridine	1
NH_3	2×10^4	Dibenz(a,h)acridine	2×10^{-1}

(2) Sawicki, E. The Chemist, XLII, #7, 259 (July 1965)

because it condenses O_2), liquid oxygen, and solid carbon dioxide.
The latter two are among the most usually used. Liquid oxygen is
hazardous to handle. No substance will be collected unless its
partial pressure in air exceeds the vapor pressure of the liquid
or solid at the condensation trap temperature, and the amount not
condensed is proportional to its vapor pressure.

Artifacts can result from condensation because reactive substances are brought together in concentrated form, which react
when increasing the temperature for sample analysis. For instance,
oxidation can be prevented when sampling organic sulfur compounds
in a CO_2 cold trap by flushing the trap with N_2 of He immediately
after sampling.

TABLE 2a.   TYPICAL COMPOSITION OF PARTICULATE
MATTER IN LOS ANGELES SMOG

Type	Component Metal	Non-metal	Percent of Total
Inorganic (70%)		H_2O	10
	$Si(SiO_2)$		10
	$Al(Al_2O_3)$		10
	Fe		10
		NO_3^-	5
	$Ti(TiO_2)$		5
	Ca		4
		$SO_4^=$	1
	Mg		1
	Ba		1
	Na		1
	K		1
		NH_3^+	0.7
	Pb		0.5
	Zn		0.5
		Cl	0.3
	$V(V_2O_5)$		0.1
	Mn		0.1
	Ni		0.1
		F	0.05
	Trace metals: Sn, Cu, Zr, B, Cr, Bi, Co		less than 0.1

TABLE 2a   (continued)

| Type | Component | | Percent of Total |
	Metal	Non-metal	
Organic (15%) (Solvent soluble)	Hydrocarbons (aliphatic:aromatic = 15:1)		
	Acids		15
	Aldehydes		
Miscellaneous (15%)	Fibers		
	Pollens (> 5 μ)		
	High Polymers		15
	Carbon		

Source: R. D. Cadle, NCAR, 1966, Ref. 83

4. Adsorption

Primarily used for purifying air rather than sample gathering
because the need for concentrating gases and vapors has decreased
by the advent of highly sensitive gas chromatographic detectors.
Non-polar substances like hydrocarbons are non-selectively collected by storage sampling. Highly polar substances which are difficult to desorb are gathered selectively by absorption or by
adsorption on specific surfaces with subsequent measurement of
changes in their electrical properties (31).

5. Sample Volume Measurements

a. Principles for measurement and calculation of gas
volumes and pressure are the most classic aspects of sample
gathering (3, 33, 34). Yet, in this application their description
is unnecessarily neglected. Sample volume measurements are therefore too often a major source of error. Sample volume measurement
is an integral part of sample gathering. The hazards are always
present of the volume measurement changing the composition of the
sample and of the sample gathering method influencing the volume
measurement.

TABLE 3.   MAJOR RADIATION ABSORBING AND IR EMITTING
SUBSTANCES IN POLLUTED AIR (18, 19)

Absorber	Region of Sharp Peaks microns	No. of Peaks	Long Path Instrumentation Ref. No.
O_2	0.687 to 0.766	2	None
O_3	0.315 to 0.330	~ 12	(20)
	IR	Several	(20)
NO_2	0.340 to 0.450	~ 50	None
NO_2	5.5 to 20	~ 8	None
NO	5.0 to 5.5	~ 2	None
SO_2	0.275 to 0.315	~ 22	None
	8 to 14	Several	(17, 20, 21)
H_2O	1.115 to 1.136	> 21	(16)
	5.0 to 5.5		
	7.1 to 10		
CO_2	2.7, 5.2, 8 to 12.0	--	(17,20)
CO	4.6	--	(17, 20, 21)
Aldehydes	3.4 to 3.9	--	None
Particles	IR & UV	non-specific	(14, 20)

        The primary standard of calibration is the spirometer.
The convenient secondary standard is the wet test meter. In some
situations an acceptable tertiary standard is a set of jealously-
guarded rotameters. Most volume calibrations should be performed
with the measuring device installed in the sampling train.

        Volume and pressure measurements are not sufficiently
reliable for generating calibrating gases, even when using care-

fully constructed and operated positive displacement pumps. For calibration an independent assay should be made. The use of stable tracer gases should receive wider application in calibration work.

TABLE 4.   SOLUBILITY OF SELECTED GASES IN
DISTILLED WATER AT 20°C

Gas	Volume Absorbed Per Volume of Water*
Nitrogen	0.015
Carbon Monoxide	0.023
Oxygen	0.031
Nitric Oxide	0.047
Ozone	0.22
Carbon Dioxide	0.878
Hydrogen Sulfide	2.582
Sulfur Dioxide	39.374

*Gas volumes reduced to 0°C and 760 torr.

The most frequently used devices provide data based on:

a-1.   Total volume samples in a measured period such as the dry or wet test meters and thus permit unrecorded variable flow rates to occur.

a-2.   Flow rates (in a measured period) which are adjustable to any desired value such a manometers and rotameters.

a-3.   Constant flow rates maintained for a measured period by critical orifices for < 30 l/min as per Figure 4 and Table 6. (23).

a-4.   Manometrically self adjusting devices for > 30 l/min (36).

a-5.   Continuous flow rate recording to permit corrections for changing flow rates. This is applicable to >> 30 l/min high volume filter sampling (35).

TABLE 5.   SELECTIVE FILTRATION SERIES —
PERCENT RETENTION (3)

		SO_2	H_2S	CH_3SH	DMS	DMDS
1.	No filter	0	0	0	0	0
2.	$NaHCO_3$	100	10	4	5	3
3.	$NaHCO_3$ and $ZnCl_2+H_3BO_3$	100	100	4	5	5
4.	$NaHCO_3$ and Ag Membrane	100	100	100	5	3
5.	$Hg(NO_3)_2$+tartaric acid and $HaHCO_3$	100	100	100	85	10
6.	$AgNO_3+H_3BO_3$+tartaric acid and $NaHCO_3$	100	100	100	100	100

*Adams, D. L.   Reference No. 56.

A major weakness in current technology is the lack of simple means for adjusting continuous and long-term (> 4 hours) sample volumes for ambient temperatue and pressure changes.

b. Techniques: generally volume measuring devices follow the sample gathering device. When the volume measurement is based on orifice metering it is essential to prevent obstruction. However, calibrations must be related to the flow rate entering the sampling train either by measurement or by correction for the pressure drop of the sampling system. The pressure drop of the calibrating device must be small compared to that of the system.

Calibrations are simplified by placing volume or rate measuring devices upstream of the sample gathering device as indicated in Figure 3. When sampling periods are short (< 60 mins) and particulare concentrations are low (< 200 μ g/m^3) it is a simple matter to clean the rotameter after several hours of use. Our evaluations indicate deposits from aerosols in sampling tubes do not interfere in the measurement of NO_x or O_3. Filtering the entering air as in Figure 5 must be done with caution to avoid interfering reactions of gases to be gathered with particles deposited. On the other hand, particles trapped in the absorber could produce an interference there. Therefore, each situation should receive appropriate consideration.

TABLE 6. PERFORMANCE OF HYPODERMIC NEEDLES AS
CRITICAL ORIFICES*

Gauge	Nominal Inside Diameter, (cm)	Flow rates (liters/min) at Needle length, (cm)		
		8.9	5.1	2.5
13		14		
15	0.14	10		9.9
17			6.9	
18			4.4	5.2
19	0.069		2.8	2.8
20			2.1	2.3
21			1.4	1.7
22			0.90	1.0
23	0.032		0.49	0.63
24				0.49
25			0.21	0.30
27	0.019			0.16
30	0.015			0.06

*760 torr, 20°C upstream

IV. ANALYSIS

Once a sample has been gathered its analysis embraces the many principles, instruments, techniques, and skills not much different from those used in analytical chemistry generally. This provides a

vast and growing range of possibilities. The analytical methods
actually selected are governed by the factors listed in Section
II.

For a variety of limits in accuracy and detectability there
are more than one method for most substances, there are no satisfactory methods for some substances and there are a multitude of
alternate methods for a few substances. In addition, there are
methods which are indices of mixed pollutants. Examples are bio-
assays such as odor, eye irritation, plant damage, and metabolic
effects on microorganisms and tissue cultures.

TABLE 7. RANGES OF HYDROCARBON VALUES EXPECTED
IN URBAN AIR MASSES

COMPONENT	RANGE, PPM IN AIR	
	Minimum	Maximum
Methane	1.2	15
Ethane	0.005	0.5
Propane	0.003	0.3
Isobutane	0.001	0.1
n-Butane	0.004	0.4
Isopentane	0.002	0.2
n-Pentane	0.002	0.2
Ethylene	0.004	0.3
Propene	0.001	0.1
Butene-1	0.000	0.02
Isobutylene	0.000	0.02
Trans-2-butene	0.000	0.01
Cis-2-butene	0.000	0.01
1,3 Butadiene	0.000	0.01
Acetylene	0.000	0.2

There exists now an Intersociety Committee for preparing a Manual of Methods for Ambient Air Sampling and Analysis. This committee is administered through the Am. Pub. Health Assn. with funds provided by the National Center for Air Pollution, U. S. Public Health Service. Methods are screened, selected and drafted by eight Substance Subcommittees on sulfur compounds, halogens, oxidants, and nitrogen compounds, carbon compounds, hydrocarbons, toxic metals, heavy metals, and radioactive substances. Promising methods are submitted for interlaboratory testing, prior to adoption as standards. This effort concerns well over 300 compounds and elements (car exhausts produce about 200 C_1 to C_{12} hydrocarbons). For the purpose of illustrating current practice and underlying principles, only a few of the most common methods will be outlined here.

A. Analysis of aerosols: Collected particles may be analyzed by determining percentages of atomic or molecular components on the entire sample or aliquots. These have been classified as bulk methods. Alternatively, the composition of individual particles can be determined. Different techniques are involved in the two cases. Any aerosol analysis seeks information about concentration, composition, and size distribution.

1. Dustfall determinations measure the amount of settleable particles, their composition, and an indication of their source.

    a. Collection

        Apparatus: 1 gal. glass jar, 4.25" wide, 10" high

        Medium: 1 liter dist. H_2O

        Duration: 5 to 30 days

    b. Analysis

        Evaporate to 500 ml

        Filter with tared quartz wool lined crucible

        Dry at 105°C, cool, weigh to yield total insoluble solids (mg)

        Ignite at 800°C, 2 hrs., cool, weigh to yield ash (mg)

        Evaporate filtrate

Dry at 105°C, 2 hrs., cool, weigh to yield soluble solids (mg)

Ignite at 800°C, 2 hrs., cool, weigh to yield soluble ash (mg)

    c. Typical values

    California: 3 to 8 gms/m^2/30 days

                 34 to 51%, soluble

                 14 to 25%, lost on ignition

    Eastern U.S.: 10 to 50 gms/m^2/30 days, industrial

                   2 to 10 gms/m^2/30 days, residential

    2. Suspended and some of the settleable particulate matter in the atmosphere is determined by collecting twenty-four hour samples on weighed 8 x 10 inch sheets of flash-fired glass fiber filters using a "high volume" sampler. After sampling the filter is reweighed to the nearest milligram to obtain the total mass collected. The filter may then be divided into sections for a variety of chemical determinations. Virtually all nonvolatile atmospheric particles down to 0.01 μ equivalent diameter arriving at the filter are collected.

    The total volume of air sampled is determined by measuring the flow rate (60 to 70 ft^3/min) at the beginning and at the end of the sampling period. As the particulate layer builds up on the filter, the flow rate decreases. Usually it is assumed that this change in flow rate occurs linearly with time. Even when this change is nonlinear, the error in volume measurement is not greater than 5% at less than 75 μg/m^3. When concentration is greater than 75 μg/m^3 the error in volume measurement is not greater than 10%. Changes in temperature, pressure and humidity are usually neglected.

    For the determination of suspended particles by weight the twenty-four hour high volume sampler is the most convenient method. Available improvements in sample design include devices (1) which maintain a constant pressure drop across the filter thus obtaining constant flow and (2) which record the decrease in flow rate with time. The high volume sampler is not suitable for obtaining hourly variations. These may be obtained by weighing smaller filters (20 to 30 mg tare) to the nearest 0.01 mg with an electrobalance. Short-term variations may be recorded continuously by total light scattering methods.

3. Size selected collections can provide information
about concentrations greater and less than 2 μ equivalent diameter.

        a. Collection

Apparatus:	As in IV 3 a, or Two-stage sampler; first-stage is a 1.3 cm diameter stainless steel cyclone collector; second-stage is a 2.54 cm diameter glass fiber filter. (Fig. 6)
Medium:	MSA type 1106 or Gelman type A glass fiber filter (40)
Rate:	18 liters/min for 4 to 8 hours
Lower Limit:	5 - 10 $\mu g/m^3$

        b. Gravimetric Analysis

Balance:	Cahn Model M-10 Electrobalance (or preferably a later model) using the 0 to 50 mg range
Sample Preparation:	All filter samples are equilibrated in a constant temperature-relative humidity chamber which approximates that of the balance room.

        c. Typical Concentrations

Clean atmosphere:	5 to 50 $\mu g/m^3$
Polluted atmosphere:	100 to 1000 $\mu g/m^3$
Second stage:	20 to 100% of total depending on weather and sources.

4. Trace Metals: More abundant metals such as Pb, Zn,
Fe in filter collections are determined routinely by atomic absorbtion or x-ray fluorescence spectroscopy. Less abundant metals such
as V, Cr, Ni, Be are determined by emission spectroscopy. In special situations for single metals colorimetric and polarographic
methods are useful.

For instance, the current situation for measuring atmospheric particulate lead concentrations is as follows: Samples are collected on lead-free filter mats. The samples are solubilized with oxidizing acids, interferences are removed by a series of extractions which result in the transfer of the lead to a chloroform solution of dithizone (sodium diphenylthiocarbazones). The color of the lead dithizonate is measured spectrophotometerically. Lead may be measured in samples containing more than 0.5 μg with a standard deviation of $\pm$ 0.1 μg. The procedure is tedious. It serves as the referee method when more convenient procedures are applied to large numbers of samples. Polarographic, emission spectrometric and atomic absorption measurements have been made on the solubilized samples. X-ray fluorescence has been measured directly on filter deposits without any sample treatment. The sensivities of these methods are compared in Table 12. The anodic stripping polarography is currently the most sensitive, and possibly among the most simple techniques. The atomic absorption techniques provide the greatest convenience, adequate sensitivity, precision and accuracy provided interferences from other constituents in the sample are eliminated by adding standardized lead solutions to the sample when calibrating.

a. Collection

Apparatus: High volume, > 200 l/min, air mover

Medium: TFA 810 cellulose sold by Staplex (38)*

Duration: 10 to 72 hours. (Increasing analytical sensitivities may soon permit collection techniques of section IV A 2.)

b. Analysis (84, 85)

Wet a section (25 to 50%) of filter with a few ml of 0.1% KOH
Dry for 1 hour at 110°C
Ash for 1 hour at 400 to 450°C in muffle furnace
Add 0.5 mo conc HCl, 5 ml H_2O, 5 drops conc HNO_3
Warm for 15 to 30 minutes
Let cool and stand overnight to precipitate silica
Filter through Whatman #40
Make up to 20 ml with H_2O

---

*Microsorban or glass fiber mats of very low trace metal content might be preferable.

Measure by atomic absorption Pb, Zn, Cd, Mn, Cu
and other solubilized elements present in the
range of the sensor and having established re-
coveries and lack of interferences from substances in this complex mixture.

To prepare samples for measurement by emission
spectrography neutralize an aliquot of the solution taken for atomic absorption.
Add a solution of 1 mg/ml pyrillidone dithio-
carbamate (absorption by ion exchange paper
appears to be an attractive alternative)
Filter through cellulose triacetate (Gelman)
Ash at 400°C (or in low temp plasma asher)
Mix 2 mg ash with 18 mg matrix consisting of
graphite and GeO_2 (Spec buffer and internal
std)

Fire at 6 amp AC arc
(or, interrupted direct current arc or plasma
arc)

c. Typical values, see references described in
    Section V, 5 and 6.

5. Organic matter is determined by the weight dissolved
in an organic solvent typically benzene. Other solvents yield
different results.

a. Collection

Apparatus: as in IV 3 a

Medium: glass fiber filters, 4" dia, or 8"
            x 11"

Duration: 1 to 3 days

b. Analysis

Determine weight of collection

Extract medium continuously for 1 hour with re-
distilled benzene (about 150 ml)

Replace the solution with fresh benzene and
continue extraction for 7 hours.

Combine the solutions and measure the volume

Remove an aliquot

Evaporate the solvent under N_2 and weigh residue, to yield the benzene-soluble organic matter.

Treat aliquots of remaining solution commensurate with determinations for specific organic compounds.

c. Typical values in California

See Table 2. Generally organic matter is about 10% of urban aerosol. Extreme meterological and source situations are in the range of 1 to 20% of urban aerosol.

6. Ring oven methods refer to a general technique which has not received as much application as it might warrant (41). It involves separation, concentration and reaction on filter paper by processes analogous to chromatography.

7. Single particle methods are used to determine size distributions, identifying sources by morphology, and micrurgy; and determining their composition by microspot tests.

a. Size distributions are strongly dependent on size parameter (4,5,6,). They can be determined by light and electron microscopy, by size-selective sampling followed by chemical analysis, or electronic sensing of particles in liquid suspensions.

Continuous monitoring of size distributions can be provided by light scattering instruments in the range of 0.5 to 10 microns projected diameter, and by combination of size selective electrostatic precipitation with condensation nuclei counting for sizes below 0.5 microns.

b. Identification of particles by recognizing microscopic morphological characteristics requires obviously special experience and a pictorial particle atlas (42). This can be extended by micrurgy, i.e. observing changes when applying physical and chemical tests and by electron microscopy (57,58,73). It is now possible to obtain elemental analyses on particles of > 1 micron with an electron (x-ray) microprobe (74).

c. Microspot tests involve collection of particles on sensitized surfaces and observing morphological and color changes. This has been used to detect gases sorbed on particles. (59,60,61,62).

d. Typically size distributions of atmospheric aerosols have < 0.5 microns mass median equivalent diameters. Composition as a function of size varies somewhat above and below that value (43). Confirmation that urban aerosol size distributions follow a power function of the product of concentration and particle diameter at any one size could considerably simplify size distribution measurements. Instrumentation measuring total light scattering is under development with this purpose in mind (44).

8. Visibility

Visibility reduction is one of the first effects of air pollution that comes to the attention of the average citizen. The gradual but continuous deterioration in visibility in urban areas is due to the increase in airborne contaminants brought about by increased activity associated with population growth.

Visibility measurements are determined by looking at prominent objects and lights located at known distances. This is a natural, quick and obvious observation to make but the results are dependent on many variables. For example, the contrast between an object and its background will determine whether or not the object will be detected by the human eye.

Two measures have commonly been used (63, 64). The meteorological range is defined as the distance at which the contrast is reduced to the point where the human eye can no longer distinguish it from the background. Meterological range is conveniently measured using instruments and takes into account scattering of light from particles observed at all angles with respect to the light source. The U. S. Weather Bureau however keeps record of prevailing visibility. This is the greatest visibility which is attained or surpassed around at least half of the horizon circle, but not necessarily in continuous sectors.

Prevailing visibility (V) in meters has been empirically related to mass concentration (M) in $\mu g/m^3$ by $V = \frac{K}{M}$ where K is a proportionality factor which has a value of about 1.3 grams per square meter $\pm$ 50%. This relationship applies only to atmospheric sampling removed from the immediate vicinity of particle emissions. On the basis of existing data, prevailing visibility is sufficiently predictable from average mass concentrations as indicated by the values in Table 7-8. To increase the reliability of this prediction, visibility observations in several areas should be taken at short time intervals at the same locations where accurate samples of suspended particulate matter are collected.

TABLE 8.   VISIBILITY AND AEROSOL MASS CONCENTRATION

Prevailing Visibility (Miles)*	Aerosol Concentration ($\mu g/m^3$)
9.2	100
4.3	200
2.8	300

*Mean values taken from Noll et al, Table III, (65)

B.   Analysis of Gases

The methods for identifying and measuring atmospheric trace gases are quite individual as compared to techniques for aerosols. This is so because gases differ widely in their physical and chemical properties. A few typical procedures are outlined below:

1. Sulfur compounds are absorbed on PbO_2 candles (9) and are converted to sulfates. These are extracted after collection, converted to a suspension of barium sulfate which is measured gravimetrically. For possible improved and simplified alternatives see references 48, 69, 70, 71.

a. Materials: Polypropylene and stainless steel assembly.
Gauze, cotton, 4" wide
PbO_2
Binder: 2.5% gum tragacanth, 12.5% EtOH, 83.0% 60°C dist. H_2O, 2.0% Dowicide B.
Slurry: 8 gm PbO_2 + 5 gm binder per candle

Candle: Coat with 13 gm slurry dry 24 hours, at room temperature at 200 torr.

b. Exposure: At least one month in California, placed into circular, louvered shelter with internal post.

     c. Analysis: Extract cloth with 5% Na_2CO_3 at 105°C, 2 hours

Cool, filter into 15 ml 6N HNO_3

Evaporate filtrate to dryness

Treat with about 1 ml 30% H_2O_2 drop by drop

Evaporate to dryness

Add 100 ml H_2O and 2 drops methyl orange (MO) indicator

Add 2 to 3 drops conc. NH_4OH, stir

Heat to near boiling, cool, filter

Acidify with 6N HCl to red MO

Heat to boiling and add hot 10% $BaCl_2$, stir

Stand overnight or 2 to 4 hours on steam bath

Filter and wash with 20 ml hot H_2O

Dry, char, and ignite at 800°C for 1 hour

Cool and weigh

     d. Calibration: None. Unexposed candle from same batch used as blank.

     e. Typical Concentrations:

California: < 0.05 to 0.4 mg SO_3/100 cm^2/day

Eastern U.S.: 0.6 to 10 mg SO_3/100 cm^2/day

    2. Sulfur dioxide is collected in a solution of sodium tetrachloromercurate to form the stable dichlorosulfitomercurate ion. Disodium ethylenediaminetetraacetate (EDIA) prevents the catalytic oxidation of SO_2 by heavy metals. After collection sulfamic acid is added to eliminate interference from nitrite ions formed by absorption of NO_2. Formaldehyde and p-rosaniline are

TABLE 9. FEATURES OF SOME CONTINUOUS AIR ANALYZERS

Pollutant	Manufacturers	Contact Column	Absorbing Reagent	Detector	Air Flow Rate L/Min	Reagent Flow Rate ml/min	Range ppm	Lower Limit ppm
Oxidant	Beckman	Helix	KI-pH 7	Colorimetric	4.0	4.0	0 to 0.7	0.01
	Technicon	Helix	KI-pH 7	Colorimetric	1.1	1.25	0 to 0.7	0.01
	Mast	Wire Helix	KI-KBr	Microcoulometry	0.140	0.021	0 to 1	0.025
	Astro Engrg.	Bubbler	KI-KBr	Electrochemical	0.20	a	0 to 1	0.001
Oxides of Nitrogen	Beckman	Spiral	azo-dye	Colorimetric	0.30	1.50	0 to 1.4	0.02
	Technicon	Spiral	azo-dye	Colorimetric	0.50	1.25	0 to 1.4	0.02
	Atlas	Spiral	azo-dye	Colorimetric	0.2 to 0.5	0.9 to 1.7	0 to 0.10 0 to 5.0	0.002
Sulfur Dioxide	Barton	Bubbler	HBr	Coulometry	0.25	a	0 to 3.8	0.03
	Leeds-Northrup (Thomas Autometer)	Helix	H_2O_2 - H_2SO_4	Conductimetric	9.4	3.33	0 to 5	0.1
	Beckman	Helix	H_2O_2 - H_2SO_4	Conductimetric	0.3 to 2	1.0 to 5.0	0 to 10.	0.02

Davis	Helix	H_2O	Conductimetric	4.0	15.0	0 to 1.8	0.01
Instrument Development	Helix	$H_2O_2 - H_2SO_4$	Conductimetric	5.20	1.7	0 to 1.7	0.02
	Bubbler	$H_2O_2 - H_2SO_4$	Conductimetric	2.35	b	0 to 1.7	0.02
Technicon	Spiral	Na_2HgCl_4 c	Colorimetric	2.1	2.0	0 to 5	0.005
Scientific Ind.	Impinger	$H_2O_2 - H_2SO_4$	Conductimetric	0.24	b	0 to 2	0.01
Calibrated Instruments	Wire Helix	$H_2O_2 - H_2SO_4$	Conductimetric	0.756	1.80	0 to 3.8	0.008
Carbon Monoxide Mine Safety Appl. Beckman	None	None	Nondispersive Infrared	not critical 1 L/min is Convenient	None	0 to 100	1
Calibrated Ins.	None	None	" " "	" " "	None	0 to 300	5
Total Hydro-Carbons Power Design	None	None	Flame Ionization	0.03	None	0 to 100	0.1
Beckman	None	None	Flame Ionization	up to 2	None	0 to 100	

a = Fixed reagent volume -- no liquid flow

b = Cumulative batch samples

c = CH_2O sulfamic acid and p-rosaniline added automatically prior to detector cell.

added to form the purple p-rosaniline methyl sulfonic acid. The
color intensity is proportional up to 2.5 μg SO_2/ml of absorbing
solution.

a. Collection

   Apparatus: Tall form extra coarse frit bubbler
                or standard midget impinger and
                suction source.

   Medium: 0.1 M sodium tetrachloromercurate
                solution containing 0.0066% EDTA,
                pH should not be less than 5.2;
                10 ml.

   Rate: Up to 5 liters/min. for a total
                volume of 30 to 40 liters.

   Lower
    Limit: 0.25 μg/ml absorbing solution and
                0.01 μl/liter of air.

b. Analysis

   Remove any precipitates by centrifugation

   When ozone is present wait 20 minutes after
   collection

   Add 1 ml of 0.2% formaldehyde in 0.6% sulfamic
   acid solution

   Mix and allow to stand 5 minutes

   Add 1 ml 0.04% specially purified pararosaniline
   hydrochloride in 2% HCl solution

   Mix and allow 20 minutes for full color development.

   Measure absorbance at 560 nm

   Calibrate with sodium metabisulfite solutions in
   Na_2HgCl_4 standardized with potassium biodate.

c. Typical Concentrations:

   California: < 0.01 to 0.20 μl/l

3. Hydrogen sulfide (56) is collected in a suspension of cadmium hydroxide and converted to CdS. Arabinogalactan added to the suspension decreases oxidation of the CdS during sampling. After collection N,N-dimethyl-p-phenylenediamine and ferric ions are added to form methylene blue. The color intensity is proportional to the sulfide concentration. The problem of absorbed sulfide oxidation when sampling air for extended periods has not yet been resolved.

    a. Collection

        Apparatus: Coarse frit bubbler and suction source

        Medium: $Cd(OH)_2$ suspension in 0.01 N. NaOH containing 1% arabinogalactan* kept dark.

        Rate: 0.05 to 1.0 liters/min for 15 to 120 minutes

        Lower
         Limit: 0.05 μl H_2S/25 ml absorbing solution and 0.01 μl/liter of air

    b. Analysis

        Reagents: 0.5% N, N-dimethyl-p-phenylene diamine hydrochloride in conc HCl, Reissner's solution (6.8% $FeCl_3$, $6H_2O$, 4.7% HNO_3) 0.1 N $Cd(Ac)_2$, 0.1 N NaOH, 0.01 N $Na_2S_2O_3$, o.01 N KIO_3, 0.75% Na_2S

        Add 1.5 ml amine solution, 0.5 ml Reissner solution
Dilute to 25 ml and shake
Develop color (blue) 30 minutes in dark
Measure absorbance at 665 nm
Calibrate with Na_2S, 2 to 16 μl H_2S equivalent, (0.08 to 0.64 μl/ml), standardizing Na_2S iodometrically immediately before use.

    c. Typical Concentrations

        California: ($<$ 0.5 to 80) x 10^{-3} μl/liter
    Community odor threshold: 6 x 10^{-3} μl/liter

_____

*STRactan 10, St. Regis Corp.

4. Mercaptans (9) are absorbed in a solution of mercuric acetate and are converted to mercaptides. After collection N, N-dimethyl-p-phenylenediamine and ferric ions are added to form a red dye. The color intensity is proportional to mercaptan concentration measured as CH_3SH.

      a. Collection

          Apparatus: Tall form, extra coarse frit bubbler; or midget impinger, coarse frit, and suction source.

          Medium: 15 ml 5% mercuric acetate

          Rate: 0.6 to 1.5 l/min. for 10 to 120 min.

          Lower
           Limit: 0.5 μl CH_3SH/15 ml absorbing solution

      b. Analysis

          Reagents: $Pb(SCH_3)_2$, N, N-dimethyl-p-phenylene-diamine hydrochloride, Reissner solution
                     (6.8% $FeCl_3 \cdot 6H_2O$, 4.7% HNO_3)

          Centrifruge absorbing solution to remove turbidity.

          Add 1.5 ml amine solution and 0.5 ml Reissner's solution, dilute to 25 ml with H_2O.

          Allow 30 minutes for color development

          Measure absorbance at 500 nm

          Calibrate with 1 to 60 μg CH_3SH added as $Pb(SCH_3)_2$ (0.04 to 2.4 CH_3SH μg/ml)

      c. Typical Concentrations:

             California: ($<$ 4 to 120) x 10^{-3} μl/l

          Community odor threshold: 10^{-2} μl/l

5. Formaldehyde is collected in a sodium bisulfite solution. After sampling, chromotropic* and sulfuric acids are added to form a colored solution. Its intensity is proportional up to

*4,5-dihydroxy-2,7-naphthalene disulfonic acid.

2 μg formaldehyde/ml.

    a. Collection

        Apparatus: Tall form extra coarse frit bubbler
                         or equivalent and suction source.

        Medium: 1% $NaHSO_3$, 10 ml

        Rate: 0.5 to 1 liter/min. up to >400
                         liters

        Lower
         Limit: 10^{-3} μl/l

    b. Analysis:

        Add 0.2 ml of 0.5% chromotropic acid* and 5 ml
        conc. H_2SO_4

        Boil at 100°C for 30 min, cool

        Dilute to 10 ml, cool

        Measure absorbance at 580 nm

        Color stability: >24 hrs.

        Calibration: iodometrically standardized solu-
        tions of sodium formaldehyde bisulfite.

    6. Carbon dioxide-containing gas is collected in inert
bags and returned to the laboratory for measurement of absorbance
in the infrared region at 4.28 μ (2330 cm^{-1}). Water vapor is removed by passing through solid absorbers.

        a. Collection: Aluminized Scotchpak or polyvinyl
        plastic bags with neoprene bulb or suitable pump
        for filling.

        Volume: 0.3 to 3 liters

        b. Analysis.

        Apparatus: Double-beam infrared spectrophoto-
                    meter equipped with 10 cm gas cells

---

*4,5-dihydroxy-2,7-napthalene disulfonic acid.

Method: Record absorption spectrum from 4.0μ
               to $4.6\ \mu$. Determine the absorb-
               ance by measuring the peak height a-
               bove the base line.

Interferences: Water vapor interference is
               minimized by passing the sample
               through Anhydrone.

Calibration: Absorbance versus volume percent
               of CO_2 in dry nitrogen.

   c. Range: 0.03% to 4%

    7. Carbon monoxide-containing gas (9) is collected in
inert bags and is transferred to the gas cell of a non-dispersive
infrared analyzer sensitized to CO. Water vapor and CO_2 are removed by solid absorbers.

   a. Collection

Apparatus: Aluminized Scotchpak plastic bags
               with neoprene bulb or pump

Volume: Several liters of air

   b. Analysis

Apparatus: Non-dispersive infrared analyzer
               sensitized to carbon monoxide.

Lower
 Limit: 0.5 to 2 μl/liter depending on length
               of sample cell

Interferences: Carbon dioxide and water vapor.
               Minimized by filter cell containing
               carbon dioxide and water vapor, tan-
               dem to sample cell; and ascarite-
               anhydrone trap preceding sample in-
               let port. These interferences may
               also be eliminated by inserting
               narrow bandpass filters which block
               radiation in the CO_2 and H_2O absorp-
               tion regions (Table 9).

Calibration: Standard gas mixtures of carbon
               monoxide in nitrogen.

c. Typical Concentrations

California: < 1 to 50 μl/liter

8. C_1 to C_5 Hydrocarbons are sampled in non-rigid containers, separated by gas chromatography (9). Each compound is
burned in a flame, creating a flux of ionized carbon atoms which
is proportional to concentration.

a. Collection

Apparatus: Aluminized Scotchpak or Teflon FEP*
plastic bags with neoprene burette
bulb or pump for filling (Fig. 3).

Volume: 1 to 5 liters

b. Analysis

Sample size: 1 ml direct, or 100 ml for conc.
by liquid O_2 freeze-out prior to
column injection

Carrier gas: Helium, 40 ml/min at 80 psig

Column temp: 0°C

Column: 2.4 m long by 3.2 mm O.D. copper tubing
17% β ,β' oxydipropionitrile on 80/100
mesh activated alumina

Detector: Flame ionization with H_2 at 27 ml/min
and O_2 at 300 ml/min

No. of cpds.
resolved: 16 out of a possible 17

Time required: 30 mins/sample including calculation of results. Less when output
is fed to a computer program, which
is recommended.

Calibration: Volumetric dilutions from 100%
gases in air

c. Typical concentrations: See Table 7.

---

*A heat sealable film of branched-chain polytetrafluorocthylene
costing about 33 cents/sq. ft.

9. Oxidants are absorbed in a neutral, phosphate buffered 1% potassium iodide solution and the concentration is reported as ozone. The liberated iodine is measured colorimetrically as
the tri-iodide ion. Theoretically, one mole of O_3 liberates one
mole of I_2. SO_2 interferes negatively mole for mole and 10% of
NO_2 appears as O_3. Other interferences are usually insignificant.
A primary standard remains to be accepted.

    a. Collection

        Apparatus: Smog bubbler and air mover

        Medium: 10 ml 1% KI buffered at pH 6.8 with
                     0.1 M KH_2PO_4 and 0.1 M Na_2HPO_4.

        Rate: 1 liter/min for 10 min

        Lower
         Limit: 0.01 ppm

    b. Analysis

        Measure absorbance at 350 nm within 10 mins

        Calibration: Standardized I_2 solutions equi-
                     valent to O_3.

    c. Typical Concentrations:

            California: Usual < 0.1 O_3 μl/liter

                     Smog 0.1 to 0.8 O_3 μl/liter

10. Nitrogen Dioxide (9) is absorbed in an azo-dye forming reagent. A stable pink color is formed within 15 minutes. Its
intensity, measured spectrophotometrically is linearly related to
the NO_2 absorbed.

    a. Collection

        Apparatus: Tall extra coarse fritted bubblers
                     and air mover

        Medium: 10 ml azo-dye reagent containing 5%
                     acetic acid, 0.5% sulfanilic acid
                     and 0.005% n(1-napthyl)ethylene-
                     diamine dihydrochloride.

        Rate: 0.4 liter/min for 10 min

Lower
Limit: 0.01 μl/liter

b. Analysis

Measure absorbance in 15 mins at 550 nm. Color
decays at rate of 3 to 5% per day at ca 25°C.*

Calibration: Standard $NaNO_2$ solutions such that
0.72 mole $NaNO_2$ is equivalent to
1 mole NO_2.

c. Typical Concentrations

California: 0.01 to 0.30 μl NO_2/liter

C. Continuous Air Analysis

1. Principles

This is achieved by any system automatically conducting analysis in closely spaced intervals 24 hours per day, seven
days per week. The disadvantages are relying too uncritically on
the data obtained; replacing a group of chemists by an equal number
of maintenance technicians, and committing a capital outlay much
greater than the situation warrants for equipment of a very limited
flexibility.

Monitoring instruments report running mean
values over periods from a few seconds (considered very fast response) to several decades of minutes. Sequential sampling followed by manual or automatic analytical techniques can achieve
similar resolution and continuity of results. In monitoring instruments using chemical techniques these are frequently simplified
in comparison to corresponding manual methods. This means the
monitoring methods are sometimes less specific and sensitive.

Careful study is recommended to determine the
initial need, to select the particular instrument, to establish
with manual methods the substances and range of concentrations to
be expected, and to organize a data reduction, storage and retrieval
capability.

Typical monitoring instruments consist of an air
mover, a transducer and an output. One text providing a detailed
overview of the operating principles is available (46). However,

---

*A medium of 0.1% sulfanilamide, 0.01% N(1-naphthylacetyl)ethylene-
diamine toluene sulfonate,0.01% polyethyleneglycol 6000 and 1.0%
$NaHSO_4$ has lower blank values and gives colors with NO_2 with decays
less than 1%/day.

many aspects applying to air analysis were not known then. These
aspects have recently been discussed by Bryan(67).

The transducers fit into five categories:

a. Physical, measuring infrared or ultraviolet absorption of the air passing through a cell without treatment.

b. Gas titration, in which a reagent gas is added
to the inlet air or the substance to be measured is removed from
the air and changes are detected in light transmission or electrical conductivity.

c. Exchange devices in which the air is brought into
contact with a reagent from which the test substance liberates
another gas which is more readily detected.

d. Electrochemical sensors in which the test substance is dissolved in a reagent causing an electrochemical property
of the solution to change.

e. Colorimetric devices in which the air with or
without addition of a reagent gas is absorbed in a liquid and a
color is formed with or without subsequent addition of another
reagent solution. The final reaction product is measured in a
flow-through colorimeter.

Some basic features of a few continuous air analyzers
in operation in many places are given in Table 9.

Despite commercial claims that have come to this
author's attention to date, all continuous air analyzers must
receive both static and dynamic calibration by the user in the
situation of use both when the instrument is initially received
and periodically thereafter. The static calibration sets the
detector performance by adjusting its response to a suitable
known standard. The dynamic calibrations set the performance of
the entire analysis system by adjusting its response to a gas
stream the concentration of which is determined independently and
at the same time it is detected by the analyzer (see section V. 1,
Figure 7).

2. Specific Substances

The current status of specific monitoring systems
for several gases is delineated below. These examples illustrate
the state of the art.

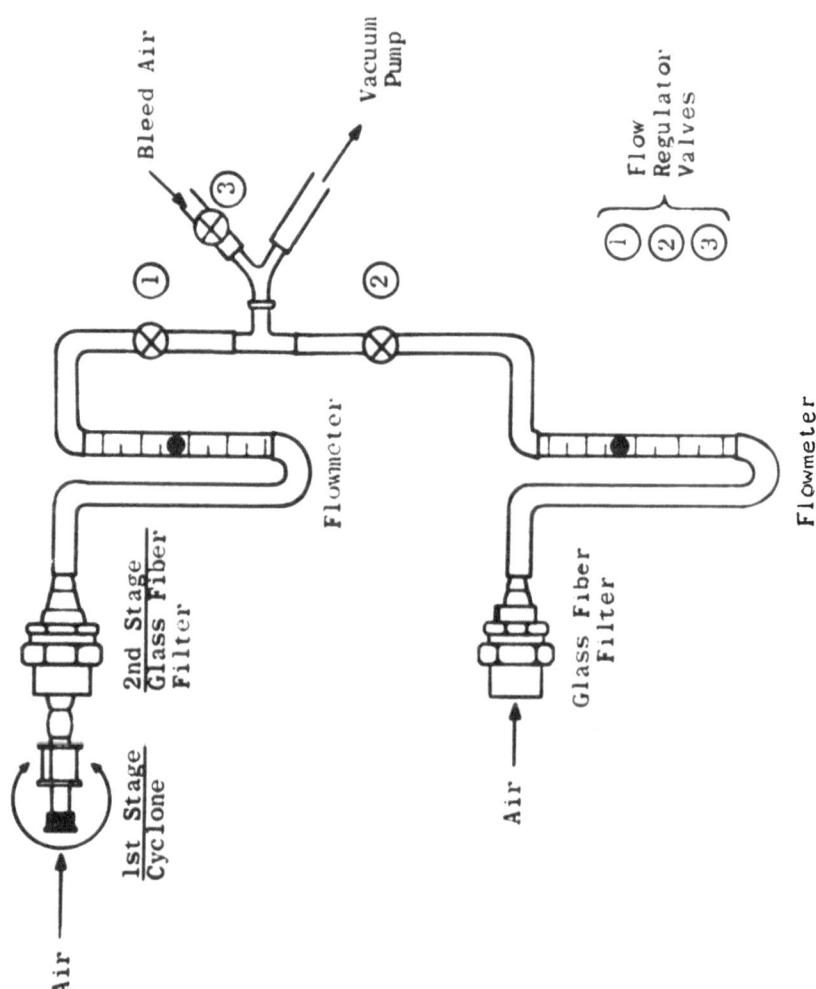

FIGURE 6. TWO-STAGE AND TOTAL PARTICLE SAMPLER

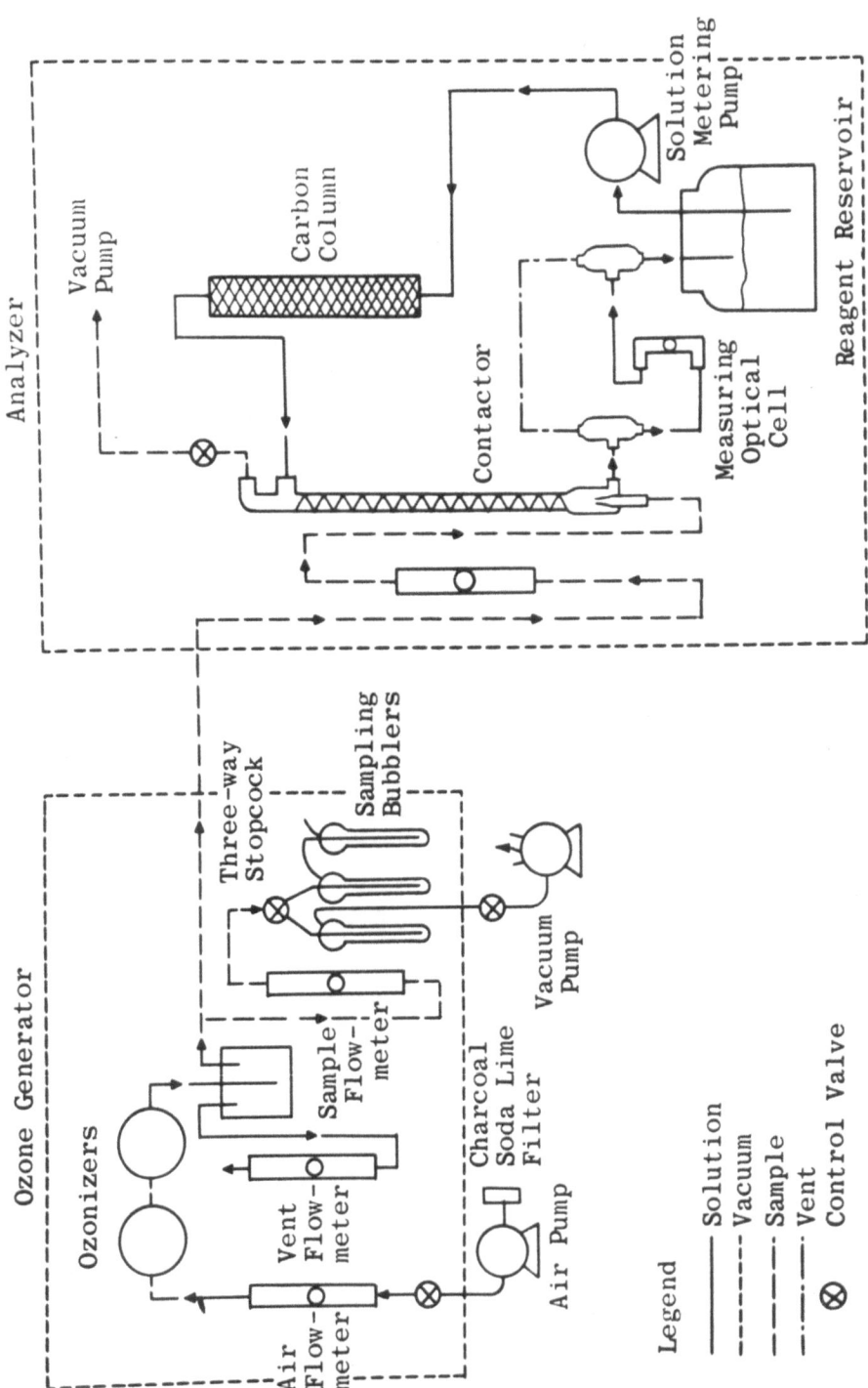

FIGURE 7. DYNAMIC CALIBRATION SETUP

a. Carbon Monoxide

Carbon monoxide is monitored continuously by measuring the absorption of nondispersed infrared radiation (NDIR) by the CO molecule. Irradiation from a hot filament is directed to a detector through two cells: a reference cell filled with dry air or nitrogen and a sample cell through which the sample air is continuously drawn. The difference in infrared absorption between the reference and sample cells creates a pressure difference in the detector. This pressure difference is transduced to CO concentrations. When the instruments are calibrated weekly with gases of known concentration based on primary standards changes from previous calibrations do not exceed 1 ppm.

The principal interferences in this method are caused by water vapor (H_2O) and carbon dioxide (CO_2). A filter cell filled with CO2 and H2O in front of the reference cell eliminates the interference from atmospheric CO_2, but only attenuates the interference from H2O. The sample is dried by passing it through a dehydrating agent which does not remove CO, such as silica gel. Los Angeles County monitoring instruments were not equipped with a dehydrating agent until 1968. CO concentrations reported by LAAPCD prior to this time are therefore somewhat higher than actual (in the range of 0 to 15 ppm depending on RH).

The NDIR system is the preferred method at this time. By equipping NDIR analyzers with narrow bandpass optical filters in the H2O and CO2 absorption region interferences from H2O can be entirely removed. This development will eliminate the need for dehydrating agents. All new equipment will probably be of this type. Current NDIR instruments for CO have a sensitivity of about 1 ppm. At concentrations less than 1 ppm the readings are unreliable. At concentrations greater than 2 ppm the error is about $\pm$ 1 ppm. More sensitive and precise instruments are under development.

b. Total Oxidant

The term "total oxidant" refers to all substances which may be present in the atmosphere that oxidize potassium iodide (KI) to form iodine (I_2) such as ozone, nitrogen dioxide (NO_2) and organic peroxides. Specific spectroscopic measurements indicate ozone is the principal oxidant in California air.

Continuous analysis of oxidant is performed by scrubbing sample air with an absorbing solution of potassium iodide (KI) buffered at pH 7. Oxidant reacts with KI to produce iodine (I_2). The concentration of I_2 is measured by a colorimeter, an amperometric cell or an electrochemical cell and recorded on a strip chart as ppm oxidant.

To establish analyzer reliability, instruments are calibrated semiannually with mixtures of ozone in air which are simultaneously measured by the analyzer and by a referee method. From 1963 to 1967 the calibrations of colorimetric type instruments were within $\pm$ 10% of the previous calibration 67% of the cases and within $\pm$ 20% in 83% of the cases.

Iodine is released from KI by nitrogen dioxide (NO_2), peroxyacetyl nitrate and other oxidizing substances. Sulfur dioxide (SO_2) in the sampled air decreases the oxidant readings by 100% of the SO_2 present. The interference from NO_2 varies with the reagent formulation. In California 2, 10 and 20% neutral buffered KI solutions are used by various monitoring agencies. The interference levels are as follows: for 2% KI, 5% of the NO_2; for 10% KI, 10 to 20% of the NO_2; and for 20% KI, 30 to 40% of the NO_2. Filters and scrubbers for removing SO_2 have been used but detailed information concerning their performance is not available. One apparently effective filter consists of glass fiber acid (H_2SO_4). But this filter is equally effective in converting NO to NO_2. When the SO_2 levels are less than 10% of the NO concentration the use of this filter is not recommended.

Consideration of existing methods indicates the colorimetric systems are still the most satisfactory. Amperometric systems appear to be subject to more interferences. Systems using electrochemical cells are very attractive because of their simplicity. However, there is no data about performance during photochemical smog conditions. Gas phase spectrometric methods utilizing infrared or ultraviolet regions appear feasible in specialized configurations. Such systems are all still under development.

c. Nitrogen Oxides

Continuous measurement of NO_2 is performed by scrubbing the sample air with an absorbing solution (Saltzman type reagent) in an automatic recording analyzer. Reaction with the NO_2 forms a red azo-dye which is measured photometrically and recorded on a strip chart as ppm NO_2.

NO is determined by first passing the air sample through an acid permanganate bubbler (oxidizer) which converts the NO to NO_2. The NO plus NO_2 originally present are recorded as NO_x. The NO concentration is then calculated by subtracting the NO_2 value from the NO_x value (parallel mode).

In Los Angeles County, NO is measured by leading effluent gas from the NO_2 absorber containing only NO to the oxidizer followed by a second NO_2 absorber. The resulting color is recorded as ppm NO (series mode).

The two modes do not necessarily yield comparable data because in the parallel mode NO_2 is lost in the oxidizer, in the series mode NO is generated from the NO_2 abosrber, and in both modes there is only partial (about 80%) conversion of NO to NO_2. Thus, in current continuous analyzers NO_2 interferes with the accurate determination of NO as a function of the NO/NO_2 concentration ratio in the sample.

Other materials found in the atmosphere generally do not interefer. However, up to 50% of the peroxyacetyl nitrate (PAN) concentration is recorded as NO_2. Although the levels of PAN are low (0.001 to 0.1 ppm), at near minimum concentrations of NO_2 and near maximum concentrations of PAN significant interference would occur.

To establish analyzer reliability, the instruments are calibrated semiannyally by procedures based on primary standards. The calibrations for NO_2 with the current type of analyzers have been within $\pm$ 10% of the previous calibration 67% of the cases and within $\pm$ 20% in 85% of the cases. Calibrations for NO have been incomplete, however, because a reliable source of NO had been lacking. This problem has now been solved.

Consideration of the existing methods indicates the continuous analyzer using azo-dye forming reagents are still the most satisfactory provided the NO_2 interference can be eliminated from the NO measurement. Removal of NO_2 (without generating NO) from the sample air with a solid scrubber consisting of chromatographic firebrick or molecular seive granules 30 to 60 mesh impregnated with triethanol amine is promising according to current tests. The RH of the existing gas is maintained at between 30 to 60% and NO is then oxidized to NO_2 in a bed of CrO_3-H_2SO_4 adsorbed on granules of chromatographic firebrick (30 mesh). Gas phase spectrometric methods in the infrared or ultraviolet regions appear feasible in specialized configurations. All systems are still under development in order to reach desirable levels of specificity, sensitivity, cost and maintainance.

d. Sulfur Dioxide

Sulfur dioxide (SO_2) concentrations are usually monitored by scrubbing sample air with a dilute solution of sulfuric acid (H_2SO_4) and hydrogen peroxide (H_2O_2). The SO_2 is converted to H_2SO_4 upon absorption in the solution thus increasing the conductivity which is recorded as SO_2 concentrations. Some conductivity analyzers use distilled water for absorption. In these interferences from CO_2 must be accounted for.

The instruments are usually calibrated annually by procedures based on primary standards. Calibrations of eight $H_2O_2-H_2SO_4$ type analyzers operated by various petrochemical industries in the Bay Area Air Pollution Control District between 1962 and 1968 have been within $\pm$ 10% of the previous calibration in 50% of the cases and within $\pm$ 20% in 85% of the cases.

Compounds other than SO_2 which dissolve in the absorber to form ions interfere in this method. The interference levels are dependent on the design of the gas scrubber and composition of the absorbing solution. Acid gases and neutral electrolytes will increase the conductivity. Basic gases (such as ammonia) and electrolytes (unless in large excess) decrease the response. The fraction of interferent appearing as SO_2 is frequently a function of the interferent-SO_2 ratio. The correction of data based on known concentrations and theory is complicated and is generally not valid. When it is known that the sum of the interfering substances is more than 10 or 20% of the SO_2 concentration, errors more than $\pm$ 10% are likely. Unfortunately, most of the interfering substances are not easily monitored. It is generally assumed that they are low with respect to SO_2 concentration.

To attain specificity the use of a colorimetric reaction which has fewer interferences would be desirable. Several continuous colorimetric instruments are commercially available. They employ various versions of the much studied West-Gaeke method. These methods all require very careful control of reagent quality. Ozone and NO_2 decrease the SO_2 reading by 4 to 20% of their concentration. The level of interference depends on their ratios with SO_2. Ozone and NO_2 are removable by scrubbing the sample air with ferrous sulfate. Systematic chemical studies are needed to find a more trouble-free dye-forming reaction.

Consideration of existing methods indicate that $H_2O_2-H_2SO_4$ conductimetric systems equipped with absorbers which minimize gas to liquid transfer of interfering substances are likely to be the most satisfactory. Gas phase spectrometric methods utilizing infrared or ultraviolet regions appear feasible in specialized configurations. Such systems are still under development.

A well-engineered system for determining total sulfur in gases is commercially available. It employs a flame photometric detector (72). It could be made specific for SO_2 by using selective adsorbents (Table 5) or a gas chromatographic column. The separation aspects are still under development.

D. Interferences

All analytical systems utilize one or several prop-

erties of the substance to be tested as a criterion. The criteria
used are not necessarily specific for a single substance. All
methods of atmospheric analysis currently in use are affected by
one or more of the compounds existing in air together with the
pollutant to be measured. There is ample evidence that changes in
procedures and in reagent formulations made for the sake of convenience or to eliminate one interference may change the interference magnitude and may even cause new interferences to arise.
These points are illustrated by the date in Table 10.

TABLE 10. EFFECT OF NO_2 OXIDANT DETERMINATION
IN ABSENCE OF O_3

NO_2 Concentration Range Tested		Conditions	Response Percent of NO_2 Conc.
μl/10 ml solution	ppm* in air		
> 0 to 114	> 0 to 36	1% KI, pH 7 bubbler	8 to 11
6.0 to 22	1.5 to 5.5	2% KI, pH 7 bubbler	6.4
8	2.0	10% KI, pH 7 contact column	21
> 0 to 4	> 0 to 1	20% KI, pH 7 contact column	30
0.96 to 1.7	0.24 to 0.43	20% KI, pH 7 contact column	12 to 47 ave: 25
8	2.0	coulombic column	10
8	2.0	U.V. photometer	2

*Based on 4-liter sample

Only very careful consideration of acceptable limits
coupled with testing can assure reliable data (46).

E. Performance Factors

For intelligent selection and application of air

monitoring instrumentation, information concerning the performance
characteristics of commercially available air analyzers is necessary. Most instrument manufacturers do not supply this information. For measuring a given pollutant, manufacturers use different
principles or the same principles in different configurations.
Users often modify existing analyzers in order to adapt the instrument to their particular needs. These differences may alter
the instrument's response to calibrating gases which is then adjusted accordingly. However, the response to interferences and variables such as temperature and relative humidity, may be significantly altered also. The generation of valid data at air monitoring stations is also strongly dependent upon design and construction which determine performance stability and durability of
components. In order to gain information concerning the performance of continuous air analyzers a systematic test program is
required. Typical performance criteria to be considered for wet
chemical colorimetric analyzers are listed in Table 11.

TABLE 11.   TYPICAL DESCRIPTION AND PERFORMANCE CRITERIA

(Wet Chemical Colorimetric Analyzers)

---

1. ANALYZER

       Model
       Manufacturer
       Vendor

2. POLLUTANT (Type of Analyzer)

3. DETECTOR

       Type
       Wavelength, nm
       Cell length, nm

4. ABOSRBER

       Type
       Concurrent
       Countercurrent
       Other

5. RANGES AVAILABLE

       a.
       b.
       c.

6. REAGENTS(S) (includes specification)

       a.
       b.
       c.
       d.

7. NOMINAL FLOW

       Airflow rate
       Reagent flow rate

8. SIGNAL OUTPUT(S)

       mv
       Current
       Telemetering

9. RECORDER

       Type, make
       Chart width

TABLE 11.   TYPICAL DESCRIPTION AND PERFORMANCE
CRITERIA (continued)

(Wet Chemical Colorimetric Analyzers)

10. NECESSARY AUXILIARY EQUIPMENT   16. DURABILITY

    a.   17. SPECIAL FEATURES
    b.
    c.   18. SUPPLIERS OF PARTS AND
                                              REAGENTS
11. POWER REQUIREMENTS
                                       19 DRIFT CHARACTERISTICS
    Volts___VAC___VDC___
    Watts___   20. DELAYS IN RESPONSE

12. PHYSICAL DIMENSIONS   21. SENSITIVITY

    Size   22. LINEARITY
    Weight
    Portable   23. REPRODUCIBILITY
    Stationary
                                       24. NOISE
13. FLOW DIAGRAM
                                       25. TEMPERATURE EFFECTS
    See Figure___
                                       26. RELATIVE HUMIDITY EFFECTS
14. REAGENT UTILIZATION
                                       27. INTERFERENCES
    Total Volume
    Single Pass
    Reusable

15. OVERALL APPRAISAL

    Quality of Construction
    Design of Layout
    Apparent Quality of Components
    Base of Maintenance

Many of the factors listed in Table 11 have sub-
categories each of which requires definition. For instance, de-
lays in response might include lag, rise, 90% response and fall
times. At present many different terms are often used synony-
mously, hwereas at other times the same terms have different meanings. No generally accepted glossary is available. Therefore,
specific meanings and definitions should be written down when
specifying performance factors.

TABLE 12.   AMOUNTS OF LEAD REQUIRED FOR MEASUREMENT
BY VARIOUS DETECTORS

Device	Lower limit of Analytical Range, μg
Emission Sepctrography	
(a) solution	20.0^1
(b) pelletization of filter	1.0^2
Atomic Absorption	
(a) flame atomization	4.0^1
(b) high temperature confined cell or volatilization	3×10^{-5}
X-ray Fluorescence	
(a) crystal energy discrimination	0.5^2
(b) electronic energy discrimination	open
Spectrophotometric	
(a) Pb dithizonate	0.5^1
(b) Pb chromate (by turbidity)	0.5^1
Color Comparator	
(a) Pb Chromate by ring oven chromatography	0.8^2
Polarography	
(a) current-voltage	5.0^1
(b) differential pulse	0.25^1
(c) anodic stripping voltammetry	0.02^1

[1] Includes dilution in transfer from a specimen into solution volumes of 5 to 15 ml.

[2] Based on 1 to 5 cm^2 filter surfaces as collected.

[3] Suitable for continuous monitoring down to 0.16 μg/m^3.

V. ANNOTATED BIBLIOGRAPHY

1. "Measurement of Air Pollutants--Guide to the Selection of
Methods", M. Kats. World Health Organization, Geneva (1969).

This is a summary of a great number of available methods for
air analysis and provides some useful perspectives on the state of
the art up to about 1960 in spite of the much later publication
date. Citations to the original literature are meticulous. However, advances in the state of the art as developed in California
during the sixties has been largely ignored.

2. "Recommended Methods in Air Pollution Measurements", Air
and Industrial Hygiene Laboratory, California State Department of
Public Health.

These are methods used primarily in California. Periodic
modifications are included as they become available. The booklet
contains both manual and continuous automatic sampling methods of
the more common pollutants e.g.: SO_2, NO_x, and total oxidant,
including methods for generating known gas mixtures and calibrating
continuous analyzers.

3. "Unified Methods for the Analysis of Pollutants in the
Free Atmosphere", S. Ube, Editor. Bull. Inst. Hyg., No. 1
Supplement, Institute of Hygiene, Prague (1966).

This is a condensed but very useful description for 17 sub-
stances: SO_2, CO, NO_2, H_2SO_4, Cl_2, H_2S, Pb, CS_2, Phenol, As, F,
NH_3, Soot, Mn, formaldehyde, and SiO_2. Alternative methods are
given for some of the substances. Sensitivity, errors, interferences, and cautionary notes given with each method are among
the most complete, quantitative and practical statements to be
found in this field.

4. "Manual of Methods of Ambient Air Sampling and Analysis",
Intersociety Committee. Am. Public Health Assn., Albany, New York;
Vol. 1 (April 1969) Vol. 2, (January 1970).

These small volumes are carefully detailed expositions of procedures for 25 substances. The procedures have all been subjected
to a multi-stage committee review process. Many of the methods are
to be accepted as standards. All are still in a "tentative" stage
because interlaboratory comparisons have not been conducted. Most
of the methods are the outcome of substantial experience in a few
laboratories in North America. For some of the substances partici-
pating societies are publishing other versions which should be com-

pared, especially the Am. Soc. for Testing Materials and the Am. Ind. Hyg. Assn.

5. "Selected Methods for the Measurement of Air Pollutants", Environmental Health Series -- Air Pollution. U. S. Dept. of Health, Education and Welfare, Public Health Service. (May 1965).

The methods for determination of common pollutants such as: SO_2, NO, NO_2, total oxidants, aliphatic aldehydes, acrolein, formaldehyde, sulfate and nitrate were selected and presented in a format covering important details and references.

6. "Methods of Measuring Air Pollution", Organization for Economic Cooperation and Development. Paris (1964).

Methods are examined critically for the measurement of grit and dust, smoke, sulfur dioxide, sulfuric acid, hydrocarbons and fluorides. Simplicity was an important criterion in selecting these procedures with some sacrifice in specificity and accuracy.

7. "ASTM Standards on Methods of Atmospheric Sampling and Analysis", American Society for Testing Materials (ASTM Committee D-22). 2nd ed. (Oct. 1962).

The subject matter covers definitions, recommended practices, conversion units, and details for 14 methods. Many of these are undergoing continuous revision.

8. "Air Pollution on Control--A Guidebook for Management", Rossano, Jr., A. T., editor, ERA, Inc., Stamford, Conn. (1969).

This is a state-of-the-art exposition geared as a text for a middle management short-course. It is oriented to source evaluation and control and includes a number of useful tabulation for emission factors, atmospheric concentrations of pollutants by city, year, and population density.

9. "Air Pollution", A. C. Stern, Vols. I and II, Academic Press, New York and London (1968).

These books are intended for professionals and are concerned with the cause, effect, transport, measurement and control of air pollution. Volume II is a state of the art exposition on most aspects of analysis, monitoring and surveying by experienced authors.

10. "Chemical Analysis of Air Pollutants", M. B. Jacobs. Interscience, New York (1960).

This is a fairly comprehensive review by an author with long experience. The procedures reflect to some extent the author's preference for classical methods. Several methods are not generally used in air sanitation laboratories.

11. "The Analytical Chemistry of Industrial Poisons, Hazards, and Solvents", M. B. Jacobs, 2nd ed. (1949).

Covers all aspects of industrial hygiene and the chemical analysis of industrial poisons. A very thorough and comprehensive textbook. Many of the principles apply, with some modification, to air sanitation.

12. "Laboratory Methods", Air Pollution Control District, County of Los Angeles, California (1958).

These are 16 methods used in the Los Angeles Air Pollution Control District laboratories. Many have been modified since publications, others are no longer in use. Three or four methods apply specifically to stack gas analysis. The phenoldisulfonic acid method for oxides of nitrogen above 50 ppm is frequently used in the Air and Industrial Hygiene Laboratory.

13. "Analysis of Air for Pollutants", J. P. Lodge and J. B. Pate, in Treatise on Analytical Chemistry: Part III, Vol. 3, I. M. Kolthoff et al, editors, Interscience Encyclopedia, Inc., New York, in manuscript 1967.

This is an excellent philosophical delineation with adequate literature documentation concerning the analytical chemistry of air sampling and analysis.

14. "Air Sampling Instruments", American Conference of Governmental Industrial Hygienists, Cincinnati, Ohio (1966).

This manual lists most of the then currently available air sampling instruments with specifications and a little performance data which help to make selections.

15. "Technical Progress Report," Los Angeles County Air Pollution Control District, Vols. I, II, and III, Los Angeles Air Pollution Control District, Los Angeles (1960, '61, and '62 respectively).

These volumes are clearly written books concerning the practice and experience of air pollution control in Los Angeles. The information is presented in a fashion which makes it suitable for general use. Volume I concerns the control of stationary sources. Volume II evaluates Los Angeles County air monitoring data and procedures for obtaining such data in the period 1951 to 1959.

And, Volume III is an excellent text concerning the detailed chemistry of photochemical smog. Each volume is printed on about 200 pages.

16. "Vehicle Emissions", Society of Automotive Engineers, Inc., Vols. 6 and 12 (1963-66).

These two volumes are compendia of the most important papers up through 1966 presented at meetings of the Society of Automotive Engineers on the composition of vehicle emissions and factors affecting them.

17. Am. Soc. Testing Materials, "Symposium on Particle Size Measurement", ASTM Publication No. 234, Philadelphia, Pa. (1959).

An excellent collection of papers each correlating theory with practice for many properties which are used for characterizing them.

18. Cadle, R. D., "Particle Size", Reinhold Publishing Corp., N. Y., (1965).

A distillation and recitation of current knowledge and classics. Important reading for grasping the scope of particle knowledge. Less helpful than some other texts. It contains a large number of empirical constants useful for making estimates.

19. Cadle, R.D., "Particles in the Atmosphere and Space", Reinhold Publishing Corp., N. Y. (1966).

Contains a very good chapter on the processes which are sources of particles in air. It provides important perspectives about the mixtures of particles one can expect to find when sampling.

20. Drinker, P. and Hatch, T., "Industrial Dust", McGraw-Hill, N. Y., 2nd ed., (1954).

A classic text of particular significance to industrial hygiene written with the authority of individuals who have devoted a life-time to such problems. The content is both descriptive and practically useful.

21. Green, H. L. and Lane, W. R., "Particulate Clouds", D. Van Nostrand, N. Y., 2nd ed., (1964).

A clear and detailed text providing sufficient explanation examples and tabulated data to permit the application of principles to practical problems.

22. Herdan, G., "Small Particle Statistics", Academic
Press, New York, 2nd rev. ed. (1960).

This is an eye-opener of how different ways of treating
particle data statistically affect results and conclusions on the
basis of simple mathematical principles. Thereby the book teaches
how to avoid pitfalls and explicitly teaches experimental design
principles.

23. Irani, R. R. and Callis, C. F., "Particle Size", John
Wiley, N. Y. (1963).

This is a practical book which teaches the measurement of
several dize distribution techniques which are applicable mostly
to powders and liquid suspensions. It spells out the limitations
and comparibility of different methods as applied to actual
examples.

24. Orr, C., Jr. and Dallavalle, J. M. "Fine Particle
Measurement", Macmillan Co., N. Y. (1959).

This book applies mostly to powders and suspensions. It
describes the principles of laboratory techniques and connects them
with physical and chemical properties of particles expressed
mathematically.

25. Spurny, K., editor "Aerosols", Czech. Acad. of Science,
Prague, (1965).

These proceedings of the Czech. 1st National Conference on
Aerosols is a record of about 100 papers on theoretical and
practical physical and chemical aspects of aerosols. The papers
appear to be carefully edited and what is unusual for such a volume, it is indexed.

26. Fuchs, N. A., "The Mechanics of Aerosols", Pergamon
Press, N. Y. (1964).

The most thorough available compendium and treatment of
aerosol theory compared with experimental data published through
1960.

27. Chapman, R. L., editor "Environmental Pollution Instrumentation", Instrument Society of America, Pennsylvania (1969).

This book is a collection of papers on air and water analysis.
The six papers on air analysis are quite general but include

meterological vehicle emission, remote sensing and sulfur dioxide instrumentation.

28. Periodicals

    a. Journal of American Industrial Hygiene Association, 14125 Prevost, Detroit, Michigan 48227.

    b. Journal of Air Pollution Control Association, 4400 Fifth Avenue, Pittsburgh, Pennsylvania.

    c. APCA Abstracts

    d. Atmospheric Environment (Formerly Air & Water Pollution Journal) Pergamon Press, 44-01 21st Street, Lond Island City, New York 11101

    e. Environmental Science and Technology, American Chemical Society, 1155 Sixteenth Street, N. W., Washington, D. C. 20036

    f. Analytical Chemistry, American Chemical Society, 1155 Sixteenth Street, N. W., Washington, D. C. 20036

    g. Chemical Abstracts, American Chemical Society, same address

    h. The Analyst, Society for Analytical Chemistry, 14 Belgrave Square, London N. W. 1, England

    i. Analytical Abstracts, Society for Analytical Chemistry, 14 Belgrave Square, London N. W. 1, England.

    j. Staub, Verlag des Vereins Deutscher Ingenieure 4 Dusseldorf 1, W. Germany

VI. REFERENCES

1. Junge, C. D., "Air Chemistry and Radioactivity," Academic Press, New York (1963).

2. Sawicki, E., The Chemist, XLII, No. 7, 259 (July 1965).

3. American Conference of Governmental Industrial Hygienists, "Air Sampling Instruments," Cincinnati, Ohio 1966.

4. Cadle, R. D., "Particle Size," Reinhold Publishing Corporation, New York (1965).

5. Green, H. L. and Lane, W. R., "Particulate Clouds: Dusts,

Smokes and Mists," D. Van Nostrand Company, Inc., New Jersey (1964)

6. Orr, C., Jr., and Dallavalle, J. M., "Fine Particle Measurement," Macmillan Company, New York (1959).

7. White, H. J., "Industrial Electrostatic Precipitation," Addison-Wesley Publishing Company, Inc., Massachusetts (1963).

8. Metronics Associates, Inc., 3201 Porter Drive, Stanford Industrial Park, Palo Alto, California.

9. State of California, Department of Public Health, Air and Industrial Hygiene Laboratory, Recommended Methods (1966).

10. Adams, D. F., and Koppe, R. K., Tappi 41, 366 (1958).

11. Gilardi, E. F., and Manganelli, R. M., J. Air Pollution Control Assoc., 13, 305 (1963).

12. Haagen-Smit, A. J., and Fox. M. M., "Cumulative Test for Atmospheric Ozone," in Rogers, Proceedings of the Conference on Chemical Reactions in Urban Atmospheres, APF Report No. 15, Air Pollution Foundation, Los Angeles, California (1956).

13. Silver-top Manufacturing Co., Inc. Pulaski Highway, White Marsh, Maryland, 21162.

14. Middleton, W. E. K., "Vision Through the Atmosphere", University of Toronto Press, Toronto, Canada (1952).

15. McCormick, R. A. and D. M. Baulch, J. A. P. C. A., 12, 492 (1962).

16. Bottema, M. et al, Annales d'Astrophysique 28, 225 (1965).

17. Law, M. J. D. and Clancy, F. K., env. Sci. & Tech. 1, 73 (1967).

18. Green, A. E. S., "The Middle Ultraviolet", John Wiley, New York (1966).

19. Leighton, Phillip, "Photochemistry of Air Pollution", Academic Press, New York (1961).

20. Dalmo-Victor Co., Inc. Belmot, California, Personal Communication (1967).

21. Barringer Research Ltd., 304 Carlingview Drive, Rexdale, Ontario, Canada, Sales literature (1967).

22. Ajax, R. L., Conlee, C. J. and Upham, J. R., J.A.P.C.A.
17, 220 (April, 1967).

23. Lodge, J. P., Jr., Pate, J. B., Ammons, B. E. and Swanson,
G. A., J. Air Pollution Control Assoc., 16, 197 (1966).

24. Schuette, F. J., "Plastic Bags for Collection of Gas
Samples", Atmospheric Environment 1, 515 (1967).

25. Vilutis and Co., So. Center Road, P. O. Box 319, Frank-
fort, Illinois, 60423.

26. Calibrated Instruments, Inc. 17 West 60 Street, New York
23, New York.

27. Fluordynamics, Greenhill Ave., and 3rd St., Wilmington,
Delaware 19805 or Sil & Associates, 1112 1/2 Maple St., Glendale,
California 91205.

28. Elkins, H. B., "The Chemistry of Industrial Toxicology",
2nd ed., Wiley, New York (1959).

29. Saltzman, B. E., Anal. Chem., 33, 1100 (1961).

30. Fracchia, M. F., Schuette, F. J. and Mueller, P. K.,
"Method for Sampling and Analyzing Carbonyl Compounds in Car
Exhaust", Env. Sci. and Technology, 1 915 (Nov. 1967).

31. Bregman, J. I. and Dravnicks, A., "Surface Effects in
Detection", Spartan Books, Inc., Washington, D. C. (1965).

32. Pierce, L., Tokiwa, Y and Nishikawa, K., "Evaluation of
Contact Columns for Nitrogen Dioxide Absorption", J.A.P.C.A. 15,
204 (May, 1965).

33. Jacobs, M. B., "The Analytical Chemistry of Industrial
Poisons, Hazards and Solvents" (1949) and "The Chemical Analysis
of Air Pollutants" (1960), Interscience, New York.

34. U. S. Dept. Health, Education and Welfare, Public Health
Service, "Atmospheric and Source Sampling", Air Pollution Training
Course Manual, Cincinnati, Ohio (1961 ff).

35. Imada, M. R. and Jeung, E., "Integrated Gas Sampling
Techniques Utilizing Critical Orifices" and Henderson, J. S. "A
Continuous-flow Recorder for the High Volume Air Sampler" both in
Proceedings 8th Conference on Methods in Air Pollution and Industrial
Hygiene Studies, State of California, Department of Public Health,
Berkeley (1967).

36. Electro-Neutronics, Inc., 22 East 8th Street, Tracy, California 95376.

37. Lodge, J. P. Contam, Control 3, 18 (1964); 4, 10 (1965).

38. Staplex, Co., 777-5th Ave., Brooklyn, N. Y.

39. Delbag Luft filter, 1 Berlin 31, Halensee, W. Germany.

40. Mine Safety Appliances Co., 201 North Braddock Ave., Pittsburgh, Pa. Gelman Inst. Co., 600 S. Wagner Road, Ann Arbor, Michigan 48106.

41. West, P. W., J.A.P.C.A. 16, 601 (Nov. 1966).

42. McCrone, W. C., "The Particle Atlas", Ann Arbor Publishers, Ann Arbor, Michigan (1967).

43. Ludwig, F. L. and Robinson, E., "Variations in the Size Distributions of Sulfur Containing Compounds in Urban Aerosols". Atmospheric Environ. 2, 13-23 (1968).

44. Lee, R. E., Jr., Patterson, R. K. and Wagman, J., "Particle-Size Distribution of Metal Components in Urban Air". Environmental Sci. and Tech. 2(4), 288-290(April, 1968).

45. Charlson, R. J., "An Instrument for Measuring the Effects of Air Pollution on Visibility". Proceedings, 8th Conf. Methods in Air Poll. & Ind. Hyg. Studies, State of California Dept. of Public Health, Berkeley (1967).

46. Siggia, "Continuous Analysis of Chemical Process Systems", John Wiley & Sons, New York (1959).

47. Mueller, P. K. et al "Chemical Interferences on Continuous Air Analysis", Proc. 7th Conf. on Methods in Air Poll. Studies, State of California Dept. of Public Health, Berkeley, Calif. (1965).

48. Huey, N. A., "The Lead Dioxide Estimation of Sulfur Dioxide Pollution". J.A.P.C.A. 18, 610 (Sept. 1968).

49. Bryd, J. F. and Phelps, A. H., "Odor and Its Measurement" in Air Pollution, A. C. Stern, Editor, Vol. II, pp 305-328, Academic Press, New York (1968).

50. Hendrickson, E. R., "Air Sampling and Quantity Measurement" in Air Pollution, A. C. Stern, editor, Vol. II, pp 3-53, Academic Press, New York (1968).

51. Calvert, S. and Workman, W., "The Efficiency of Small Gas Absorbers", Amer. Ind. Hyg. Assoc. J. 4, 318 (August 1961).

52. Calvert, S. and Workman, W., "Estimation of Efficiency for Bubble-type Gas Absorbers". Talanta 4, 89 (1960).

53. Gage, J. C., "The Efficiency of Absorbers in Industrial Hygiene Air Analysis". Analyst, 85, 196 (1960).

54. Lockhart, F. J., "Fundamentals of Mass Transfer as Applied to Gas Analysis Instrumentation". Proc. 9th Conf. Methods in Air Poll. & Ind. Hyg. Studies, State of California, Department Public Health, Berkeley, California (1968).

55. Snyder, L. J. and Henderson, S. R. "A New Field Method for the Determination of Organolead Compounds in Air", Anal. Chem. 33, 1175 (August 1961).

56. Adams, D. F. et al "Analysis of Malodorous Sulfur-Containing Gases", Proc. 9th Conf. Methods in Air Poll. and Ind. Hyg. Studies, State of California Dept. of Public Health, Berkeley, California (1968).

57. Waller, R. E., "an Electron Microscope Study of Particles in Town Air," Int. J. Air Water Poll. 7, 779 (1963).

58. Frank. E. R. and Lodge, J. P., "Morphological Identification of Airborne Particles with the Electron Microscope,"J. de Microscopie 6, 449 (1967).

59. Cadle, R. D. et al, "Identification of Particles in Los Angeles Smog by Optical and Electron Microscopy", Arch. Ind. Hyg. Occ. Med. 2, 698 (Dec. 1950).

60. Lodge, J. P., "Analysis of Micron-Sized Particles" Anal. Chem. 26, 1829 (Nov. 1954).

61. Waller, R. E., "Acid Droplets in Town Air", J. Air Water Poll. 7, 773 (Oct. 1963).

62. Green, W. D., "Sensitized Films for the Detection and Analysis of Chemical Aerosols", Proc. 9th Conf. Methods in Air Poll and Ind. Hyg. Studies State of California Dept. of Public Health, Berkeley, Calif. (1968).

63. Robinson, R. "Effect on the Physical Properties of the Atmosphere" in Air Pollution, A. C. Stern, editor, Vol I, pp. 349-400. Academic Press, New York (1968).

64. Middleton, W. E. K, Vision Through the Atmosphere, University of Toronto Press, Canada (1963).

65. Noll, K. E., Mueller, P. K., and Imada, M., "Visibility and Aerosol Concentration in Urban Air", J. Atmos. Environ. 2 (Oct. 1968).

66. Charlson, R. J., Horvath, H., and Pueschel, R. F., "The Direct Measurement of Atmospheric Light Scattering Coefficient for Studies of Visibility and Pollution", J. Atmos. Environ. 2 (Oct. 1968).

67. Bryan, R. J., "Air Quality Monitoring" in Air Pollution, A. C. Stern, Editor, Vol. II, pp 425-464. Academic Press, New York (1968).

68. Pierce, L. B. et al "Validation of Calibration Techniques", Proc. 8th Conf. Methods in Air Poll. and Ind. Hyg. Studies, State of California Dept. of Public Health, Berkeley, Calif. (1967).

69. Huey, N. A., Waller, M. A., and Robson, C. D., "Field Evaluation of an Improved Sulfation Measurement System", paper no. 69-133, 62nd Annual Meeting of APCA, June 22-26, 1969.

70. Fukui, Kanno, "The Alkaline Filter Paper Method for Measuring Sulfur Oxides, Nitrogen Dioxide and Chloride in the Atmosphere", Proc. of the International Clean Air Conference, Part I, London 4-7, p. 231, 1966.

71. Waller, M. A. and Huey, N. A., "Evaluation of a Static Monitor of the Atmospheric Activity of Sulfur Oxides, Nitrogen Dioxide and Chloride", paper no. 69-90, 62nd Annual Meeting of APCA, June 22-26, 1969.

72. Melpar, 7700 Arlington Boulevard, Falls Church, Va. 22046

73. Heard, M. J. and Wiffen, R. D., "Electron Microscopy of Natural Aerosols and the Identification of Particulate Ammonium Sulphate", Atmos. Environ. 3:3, 337-340 (May 1969).

74. Bayard, Michael, "Microprobes", American Laboratory (Jan. 1969).

75. Umbraco, R. A., "Fluorocarbon Films for Sampling Bags", AIHL Report No. 80, State of California, Dept. of Public Health, Berkeley, (March 1970).

76. Rossano, Jr., A. T., editor, "Air Pollution Control -- A Guidebook for Management". ERA, Inc., Stamford, Conn.(1969).

77. Morgan, G. B. and Ozolins, Guntis, "Air Quality Surveillance", Proc. 11th Conference on Methods in Air Pollution and Industrial Hygiene Studies, University of California, Berkeley (1970).

78. Saltzman, B. E., "Factors in Air Monitoring Network Design", Proc. 11th Conference on Methods in Air Pollution and Industrial Hygiene Studies, University of California, Berkeley (1970).

78. Percy, R. B. and Hildebrandt, P. W., "Air Monitoring Criteria", Proc. 11th Conference on Methods in Air Pollution and Industrial Hygiene Studies, University of California, Berkeley (1970).

80. Bell, Gordon, "Commentary on Monitoring Site Selection", Proc. 11th Conference on Methods in Air Pollution and Industrial Hygiene Studies, University of California, Berkeley (1970).

81. McGuire, Terry and Noll, K. E., "Relationships Between Concentrations of Atmospheric Pollutants and Averaging Time", Proc. 11th Conference on Methods in Air Pollution and Industrial Hygiene Studies, University of California, Berkeley (1970).

82. Whitby, K. T. and Liu, B. Y. U., "Atmospheric Particulate Data--What Does It Tell Us About Air Pollution?", Proc. 11th Conference on Methods in Air Pollution and Industrial Hygiene Studies, University of California, Berkeley (1970).

83. Cadle, R. D., "Particles in the Atmosphere and Space", Reinhold, New York (1966).

84. Thompson, R. J., Morgan, G. B. and Purdue, L. J., "Analysis of Selected Elements in Atmospheric Particulate Matter by Atomic Absorption". In Analysis Instrumentation, Vol. 7, Instrument Society of America, 530 William Penn Place, Pittsburgh, Pa., 15219 (May 1969).

85. Bay Area Air Pollution Control District, Private Communication (1970).

ACKNOWLEDGEMENTS

A broad ranging summary of this kind, I can prepare only with the dedicated but critical support of my colleagues at the Air and Industrial Hygiene Laboratory. I am especially grateful to Mrs. Shirley Tocchini for preparing the typescript and Messrs. Yoshiro Tokiwa, Ronald Stanley and Louis Pierce for assisting with the text.

METEOROLOGY

James G. Edinger

Professor of Meteorology

University of California, Los Angeles

The Role of Meteorology in Air Pollution

Atmospheric conditions not only specify the path taken by the air from pollution source to affected target but determine the extent to which the pollutant becomes diluted or chemically altered by the time it reaches the target. The atmosphere's role as transporter of pollution is obvious but perhaps the wide variation in its power to dilute is not so well appreciated. Reflect for a moment on the fact that on a day when a large metropolitan area is enjoying sparkling clear air, unlimited visibility, deep blue sky-- on that day there probably is just as much aerial contamination being injected into the atmosphere as on the most smoggy day on record. This vast difference in dilution can be ascribed simply to a difference in atmospheric configuration. Our purpose here is to examine the basic relationships between the structure and motion of the lowest layers of the atmosphere and the transport and dilution of introduced contaminants.

Atmospheric Dilution of Pollutants

The dilution of a cloud of pollution is a mixing of the cloud with its atmospheric environment. Two different sorts of mixing or diffusion take place simultaneously: molecular diffusion and eddy diffusion. The rate of transfer of material across the cloud or plume boundary by molecular motions is very small compared with that which is generated by "fluid" or eddy motions. Even the smallest of these eddies transfer material at rates orders of magnitude greater than does molecular diffusion. So we can neglect the molecular effects and concern ourselves only with the eddying motion of the atmosphere. But this encompasses a very wide range

of eddy sizes--everything from the small wisp traced out by the
smoke from a cigarette to the large whirls, perhaps a thousand
miles in diameter, seen in satellite cloud photographs. And the
diffusion rates stemming from these widely disparate scales are
themselves grossly different, taking on values that range over more
than ten orders of magnitude when computed for the whole spectrum
of "fluid" motions in the atmosphere. Clouds of pollutants that
are injected into the atmosphere are observed to disperse and
diffuse at rates that increase markedly as the clouds themselves
increase in size since the scale of mixing motions to which they
are subject increases also with size. Another way of phrasing it:
It is the atmospheric motion on a scale small compared to the
cloud that actively mixes the cloud with its environment. Motions
of a scale large compared to the cloud only transport the cloud.
Motions of a scale comparable to the cloud just distort or deform
the cloud. So to follow a cloud through a long history from small
concentrated volume to very large and dilute volume, of say con-
tinental dimensions, we observe a continuing enhancement of the
dilution mechanism as progressively each larger and more effective
scale of eddy mixing motions joins those already diluting the expanding pollutant cloud.

        Fortunately we are concerned here with a relatively short in-
terval in the history of a pollution cloud during which the size
of the cloud does not change a great deal. We will focus our attention on the pall of contaminants that comes into the atmosphere
from a large metropolitan complex. The cloud at inception is not
a small point source but a myriad of individual sources densely
distributed over what we may think of as a single large area source
measured in terms of hundreds of square miles. Furthermore, it is
not an instantaneous source, like for instance an explosion, but
is a continuous though variable one. The cloud it forms usually
takes the form of a murky wake downwind of the city. On the days
of most interest to us, when dilution is extremely low and pollution
concentrations high, the width of the plume or wake will increase
very little with time, remaining the same order of magnitude as
the width of the source, the diameter of the metropolitan complex.
So the mixing motions doing the diluting do not increase in scale
significantly and our problem is simplified.

        The stirring and mixing motions that dilute a large city's
contaminants are too small to be represented on the usual weather
map on which the data points are separated by 50 or 100 miles.
It takes a very dense network of wind reporting stations just to
represent the flow patterns that are doing the distorting and deformation of the polluted pall. Los Angeles has such a mesometeor-
ological network, about 50 stations spaced at distances of around
5 miles or so. Streamline analyses of this data sometimes deline-
ate eddies and whirls 10 or 15 miles in diameter. (See Fig 1)

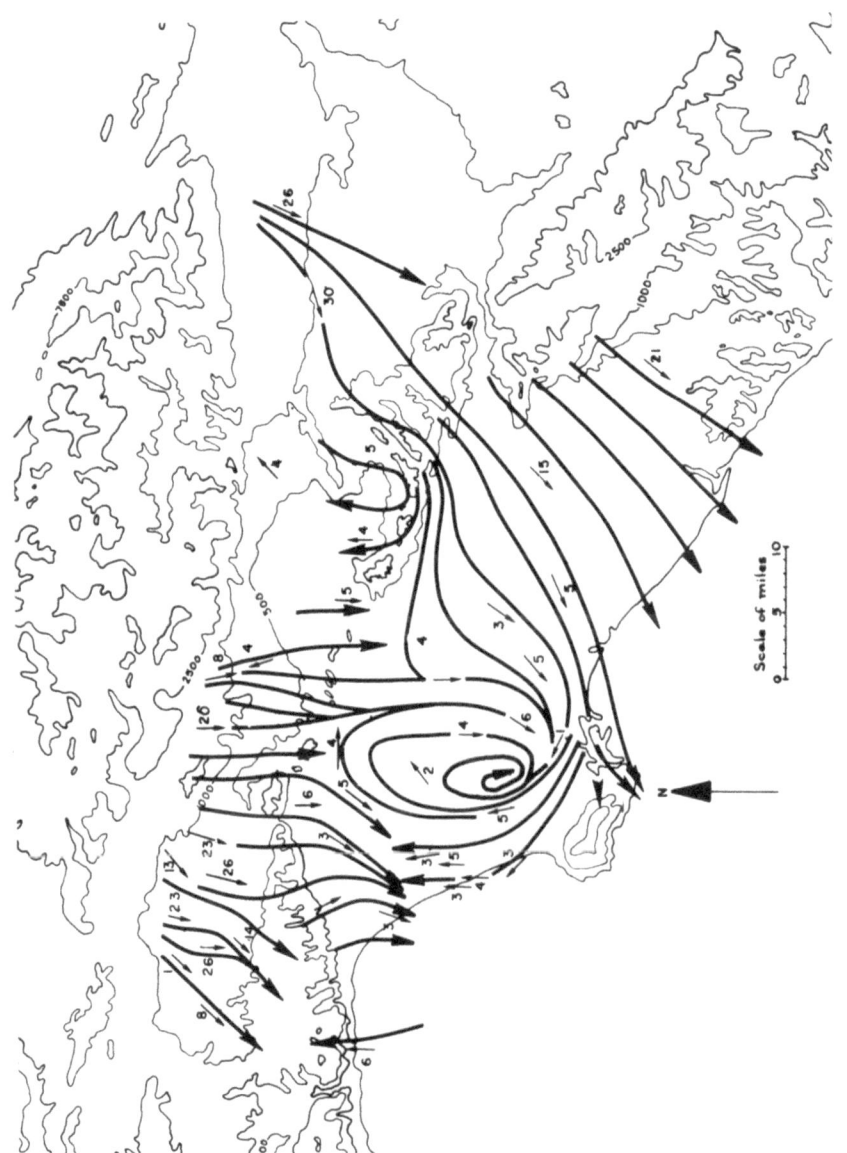

Fig. 1. Surface wind pattern over the Los Angeles Basin, 0500 PST, Nov. 22, 1957.

But the smaller eddies responsible for the dilution are so small
that they never affect more than one station at a time, and so go
completely undescribed. They are detectable, however, in another
type of wind record, the continuous wind speed and direction traces
of the individual anemometer and wind vane. These show the apparently random fluctuations of the wind at a single station as the
atmosphere's wide spectrum of eddies drift by. Typically we must
depend upon statistical descriptions of this turbulent flow, root
mean square deviations about the mean speed, as an example. Descriptors of this sort are used by the practitioner in determing diffusion
rates. A wide variety of formulae using statistical variables have
been derived in efforts to suit the multitude of configurations of
both source and atmosphere. It would be inappropriate here to go
into the details. Instead it may be more helpful to examine the
causes behind these mixing motions so that one might infer their
existence from other properties of the atmosphere that are measured
routinely.

The lowest part of the atmosphere is subject to a number of
forces that give rise to the motions that we observe there. And
these forces do not operate equally in the horizontal and the vertical, gravity being the most obvious example. It is not surprising
that air motions show some directional preferences and that the dilution rates that depend on air motions also differ in the horizontal and vertical. It is convenient, therefore, to separate our
discussion in two parts. First we consider what it is that brings
about the motions that mix air contaminants vertically.

Vertical Mixing of Pollutants

If one parcel of air is rising relative to a neighboring parcel,
one of two effects may be responsible. The parcel may have encountered an obstacle and been deflected upward or it may somehow or
other have become buoyant. The sort of mixing motions that results
when air moves over rough terrain, mechanical mixing, is pretty
well confined to the layer next to the ground. Buoyancy effects on
the other hand are not confined to the lower boundary of the atmosphere and are the basic cause of most vertical mixing throughout
the depth of the lower atmosphere. This may seem odd at first when
we consider that the major source of heat for the atmosphere is
the surface of the earth. But buoyancy forces can be developed in
mid-air, so to speak, without any transfer of heat to or from an
air parcel. As an example, let's consider a layer of air which,
for simplicity's sake is isothermal, has the same temperature at
all heights. Select one small parcel of air interior to the layer
and lift it 100 meters through the layer. Its environment at its
new location has the same temperature as its original environment,
but the parcel's temperature has decreased by expansion while
moving upward to a position of lower pressure. If we have succeeded

in preventing any heat exchanges between the parcel and its environment while lifting it, it will have cooled at the dry adiabatic rate, 1 deg C per 100 m, by virtue of the work it did in expanding against its environment. Being one degree cooler than its new environment it will at this position be negatively buoyant and if released will return to its original level. We can make the parcel positively buoyant by displacing it in the opposite direction, downward. 100 meters below its original level it will be one degree warmer than its environment. If released it again seeks its original level. We must conclude then that an isothermal layer is stable with respect to vertical displacements. By that we mean that all vertical mixing motions both up and down will be opposed by restoring buoyancy forces. In a stable layer, then, diffusion rates will be opposed by restoring buoyancy forces. In a stable layer, then, diffusion rates will be small.

Usually the lowest layer of the atmosphere is not isothermal. Typically it is stratified in such a way that its temperature at one level is about 2/3 deg C cooler than it is 100 m below it, or in other words it has a lapse rate of temperature something like 2/3 C per 100 m rise. If we repeat the above experiment on a day when the layer in which we are interested happens to have this typical lapse rate, we find that this layer also is stable. 100 m above its original level a displaced parcel again is cooler than its new environment, but this time only by 1/3 deg C. And displaced 100 m below it is 1/3 deg warmer. Once more we have a restoring force, but a much smaller one. So we have a stable layer, but not nearly so stable as an isothermal layer and one not nearly so averse to vertical mixing motions.

If an atmospheric layer just happened to have a lapse rate numerically equal to the dry adiabatic rate of cooling (1°C per 100 m), vertical displacements within it would not develop any buoyancy forces at all. Such an atmospheric layer is said to have neutral stability. Since vertical motions go unopposed by buoyancy forces diffusion rates are high.

But in some layers the lapse rate is even higher. Consider the layer whose lapse rate exceeds the dry adiabatic rate of cooling, for example 1 1/3 deg C decrease per 100 m. An upward displacement of 100 m lends a parcel 1/3 deg warmer than its environment. No restoring force. Instead there is a force in the direction of the displacement. And the further the air parcel moves from its initial position the greater the buoyancy force in that direction. This layer is unstable to vertical displacements. Unstable layers have high diffusion rates. And the greater the lapse rate the greater the diffusion rate.

To summarize then, the property of atmospheric layers that pre-

scribes how rapidly contaminants may diffuse through them and become diluted by them is their vertical temperature gradients. And
these are classified as stable, neutral, or unstable in order of
increasing potential for diffusion pollutants. (See Fig. 2)

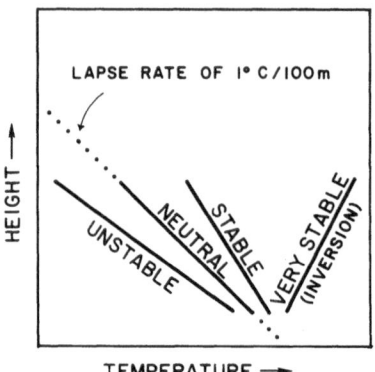

Fig. 2. Stability of atmospheric layers as a function of their
         lapse rates (vertical temperature gradients).

     Of most interest to us is the most stable configuration, the
inversion, in which temperature actually increases with height.
Inversion layers are so successful at suppressing vertical mixing
that in most cases they effectively halt completely the vertical
transfers of pollution by eddy diffusion.

Horizontal Transport and Diffusion of Pollutants

     Now to establish the role that horizontal air motions play in
the dilution of aerial contaminants let's consider a large metropolitan area on a day when a low, strong inversion exists over it.
The strong inversion is so stable that it effectively damps out all
vertical mixing motions at its level. And although vigorous mixing
motions may exist in the less stable, perhaps even unstable layer
between it and the ground, the inversion puts an upper limit on
eddy diffusion. Pollution introduced at the ground becomes diluted
with the sub-inversion air, but the absence of mixing motions in
the inversion layer itself prevents any further penetration of the
contaminants in the vertical.

     In the absence of any horizontal motion then we would have a
large area source continually pouring pollution into a fixed volume
delineated horizontally by the periphery of the source area and
vertically by the base of the inversion layer. So long as conditions remained unchanged pollutant concentrations would continue
to mount in this fixed volume of air. Now introduce some wind.
The most obvious and most important effect of horizontal air move-

ment is to bring new, unpolluted air into the source area across
its upwind boundary and at the same time remove an equal volume of
polluted air on the downwind side of the city. Any such net transport of air across the source area puts an end to the unlimited increase in pollutant concentration in the cloud. For any given wind
speed and time interval there is a prescribed volume of air into
which the area source injects contaminants. Double that wind speed
and you double that volume and as a consequence halve the pollutant
concentration. The analogy with a constant flow of sewage into
a stagnant pond, a slow moving river, or a fast flowing river is
quite close. The major point of difference between the air flow
and water flow is that the polluted air does not have such well
defined lateral boundaries. The river banks provide the outer boundary for water pollution. The "rive" of polluted air typically is
able to widen under the influence of horizontal eddy diffusion.
This additional dilution, however, can usually be considered secondary to that provided by bulk horizontal transport of new air past
the source.

So, for a large city we have a rather simple picture for that
most important pollution situation, the day with low strong temperature inversion. The cloud of pollution produced by the multi-
tidue of individual sources, some fixed, some moving, some continuous, others intermittent is one fairly continuous pall over the
city stretching out downwind of it with a sharply defined upper
boundary. Knowing the mean wind speed, the height of the base of
the inversion layer, the diameter of the source area, and the pollutant emission rate one can make a first approximation of the mean
concentration of contaminants in the cloud. Here we make the un-
warranted assumption that nothing is happening to the contaminants
that would remove or alter them either chemically or mechanically.
We also oversimplify the horizontal motions of the air. The local
terrain typically impresses important diurnal fluctuations on air
motions and a variety of other constraints that complicate the
horizontal and vertical flow field. These can lead to important
distortions of the above simple picture and must be considered.

The most obvious effect that terrain has on horizontal air
motion is to steer and channel the air flow, much as mountain ranges
prescribe the courses of rivers. But the analogy is far from perfect. Just how good it is depends on the stability or instability
of the atmospheric layer in question. If the layer of air covering
the terrain has an unstable lapse rate, air encountering a mountain
will find it much easier to move up and over the obstacle than to
deviate its course horizontally, positive vertical displacements
giving rise to positive buoyancy forces. But we are not very concerned here with unstable layers since they allow the rapid dispersal and dilution of pollutants vertically and present no pollution problem. Our attention is focused on the very stable layer

with the concentrated load of contaminants confined between it and
the ground. Air within such a layer develops opposing buoyancy
forces when displaced vertically and behaves very much like a river
in mountainous terrain.

It is popular to ascribe the accumulation of smog in the Los
Angeles Basin to this effect: "the damming or trapping of the smog
by the encircling mountain ranges." But this cliche resembles
reality only part of the time, the nighttime. Then the mountains
do act like dams--steer the flow. But in the daytime their role
is ambivalent. To understand why we must abandon the adiabatic assumption, at least for that part of the smoggy layer that is in
contact with the ground. The adiabatic assumption is a remarkably
good assumption for air out of contact with heated or cooled terrain.
Other heating effects such as radiation that can operate in the free
atmosphere act too slowly to appreciably influence the buoyancy
forces brought about by small scale adiabatic vertical displacements.
But the heating and cooling of air in contact with mountain slopes
is a large effect and cannot be ignored. At night it serves to
make the very lowest layers even more stable which enhances the
effectiveness of the mountains as steerers of the flow. But in the
daytime the layer of air next to the mountain, being heated from
below becomes unstable, in fact positively buoyant with respect to
the air at the same level out away from the slopes, so it moves up-
slope, carrying its pollution with it, just like smoke up the chimney. Far from being a dam the slopes become a vent for the polluted
lower layer during the day. It is not uncommon for the smog to in-
vade Mt. Wilson, almost 6000 feet above Los Angeles, when the smog
layer over the city itself is only about 1000 feet deep. It is
air pollution in transit up the chimney about to be released into
the upper layers where it rapidly mixes with a deep layer of air
free of inversion layers. So in the daytime we have the picture
of a broad shallow river of air pollution, the bulk of which out
away from the mountains moves as if steered by the ranges and
channeled by the passes. But at the edges of the cloud next to
the heated slopes, in very un-riverlike fashion, the pollution is
being drawn off uphill and vented above the ridges.

Terrain also plays a part in determining the rate at which
fresh air is brought in over the source. Winds typically are not
constant but have a diurnal fluctuation, a cyclic variation geared
to the coming and going of the sun and instigated by differences
in heating and cooling of different terrain surfaces. Undoubtedly
the largest inhomogeneity in terrain as regards temperature variations is a coastline. Temperature contrasts across a coastline
during a clear day set in motion a large convection regime that
produces at ground level a sea-breeze (wind directed from sea toward land). At night the temperature gradient is just reversed
resulting in a land-breeze in the opposite direction. Normally

this fluctuating wind is superposed on the large scale flow provided by pressure patterns seen on conventional weather maps. The
result is a perturbation of this general flow. In Los Angeles in
the summer, for instance, the general flow is from sea toward land.
During the day the sea-breeze enhances this flow and at night the
land-breeze opposed it. The result is on-shore movement of around
10 mph during the warm parts of the day and very nearly calm or
very weak land-breeze at night. Figure 3 shows the halting paths
taken by two arbitrarily chosen parcels of air as they moved across
the Los Angeles Basin. Note that the afternoon hours are hours of
good ventilation but the early morning is a period of near stagna-
tion. All other things being equal (which they aren't) the source
area should have a more dilute polluted wake in the afternoon under
the benevolent influence of the sea-breeze than it does in the
early morning hours when virtually no new air is brought in over
the source area and one could expect an accumulation of contaminants
that stops only with the mid-morning onset of the sea-breeze.

Diurnal variations in ventilation can also occur in the absence
of coastlines, for example, valley-mountain regimes. The thermal
contrast producing these winds is that between the alternately
heated and cooled slopes of mountains and the air at the same level
out over and well above the valley floor--air that is scarcely
heated or cooled at all. Figure 4 shows these daytime up-valley
and up-slope flows and nighttime down-valley and down-slope winds
for a valley in western Argentina. In those important instances
where the large scale pressure patterns prescribe weak flow it is
these diurnal wind fluctuations that provide whatever horizontal
ventilation an area receives. It is apparent that each locality
with its unique terrain has its own varying potential for receiving
new, unpolluted air depending on both general wind conditions and
locally generated winds.

Conditions for Inversion Formation

A combination of large scale and local weather conditions also
determine when and where inversions will form. A variety of weather
situations can produce low level temperature inversions. They
furnish either cooling from below or heating from above or both
for the lower part of the atmosphere.

Cooling from below usually comes about by nighttime radiation
losses at and near the ground. These are most pronounced when the
sky is clear (no radiation back to the ground from the clouds) and
when the wind is light (a minimum of mechanical mixing transferring
heat from warm air above to cold air below). Under such conditions
very marked inversion layers, usually a few hundred feet in depth,
develop right at the surface. Figure 5 shows just how a ground
inversion (one based at ground level) developed on a particular

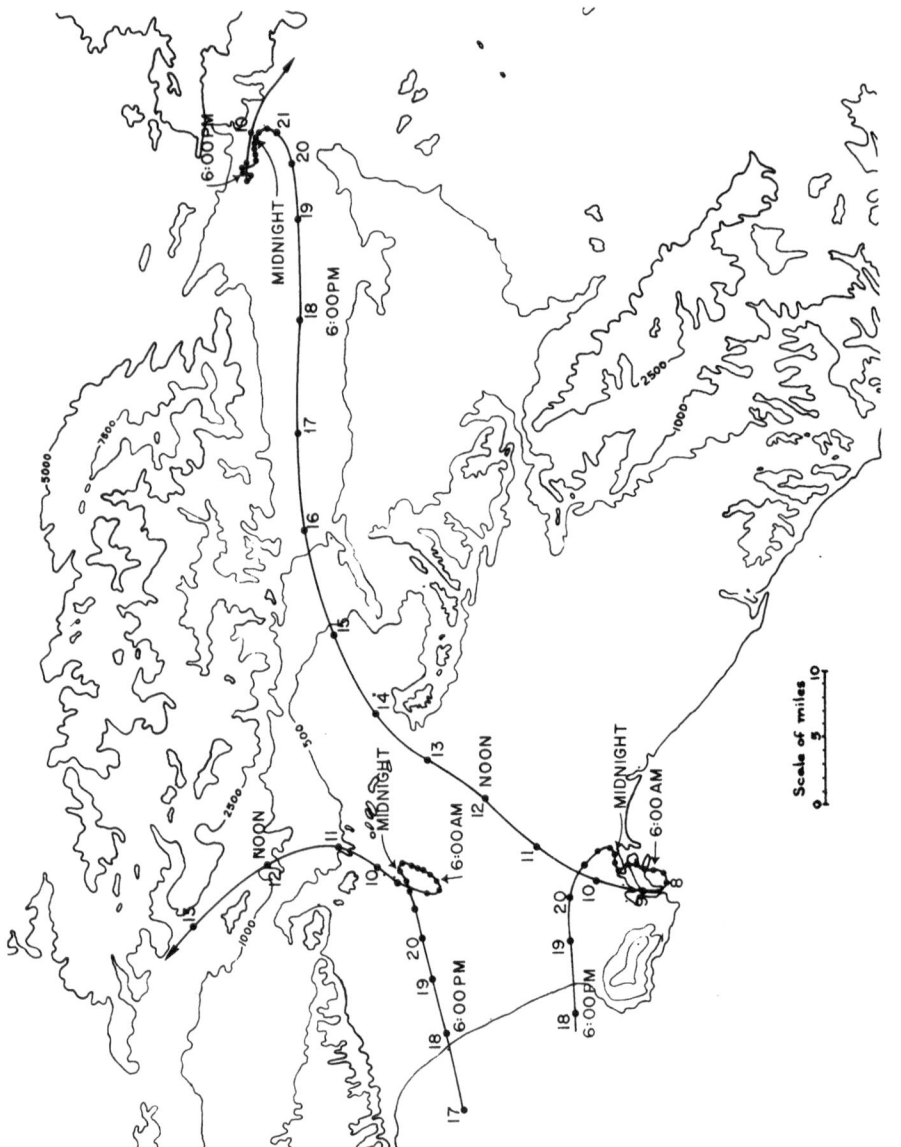

Fig. 3. Typical summer trajectories across the Los Angeles Basin.

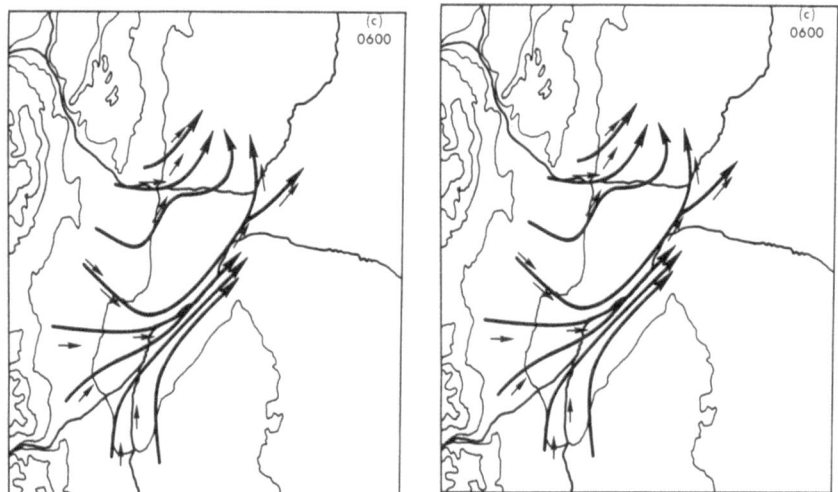

Fig. 4. Prevailing wind patterns for 6 a.m. and 12 noon in the
Tunuyan Valley, Argentina.

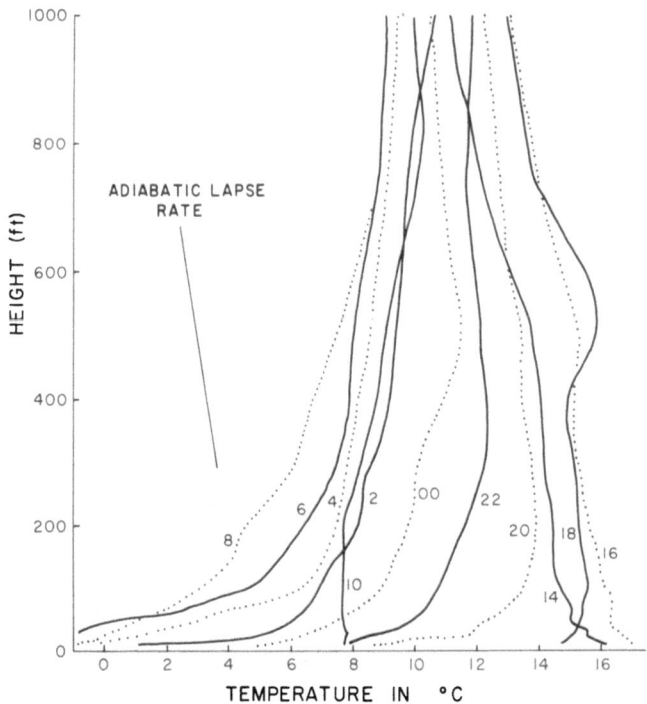

Fig. 5. Variations of vertical temperature stratification with time
time of day (MST) at Gila Bend, Arizona, Feb. 5-6, 1946.

night at Gila Bend, Arizona, under a condition of weak winds and
a clear sky. Inversions that form in this way are called radiation
inversions. Note in the figure that the inversion disappears during
the daylight hours when the situation is reversed and the air is
heated from below.

Another way in which air can be cooled from the bottom is to
move it horizontally from a warm lower boundary to a cool one, for
example from the heated daytime land surface across the coastline
to a cool ocean. Eddy diffusion transports heat from the air to
the ocean, cooling the lowest part of the atmosphere. Figure 6
shows the results of this process on warm air streaming from the
Los Angeles Basin out over Santa Monica Bay during a "Santa Ana"
wind.

Heating a layer from above invokes less obvious processes.
Only one, adiabatic compression, is of importance. It comes about
when a layer of the atmosphere becomes shallower. For example, if
a volume of air next to the ground is spreading out horizontally,

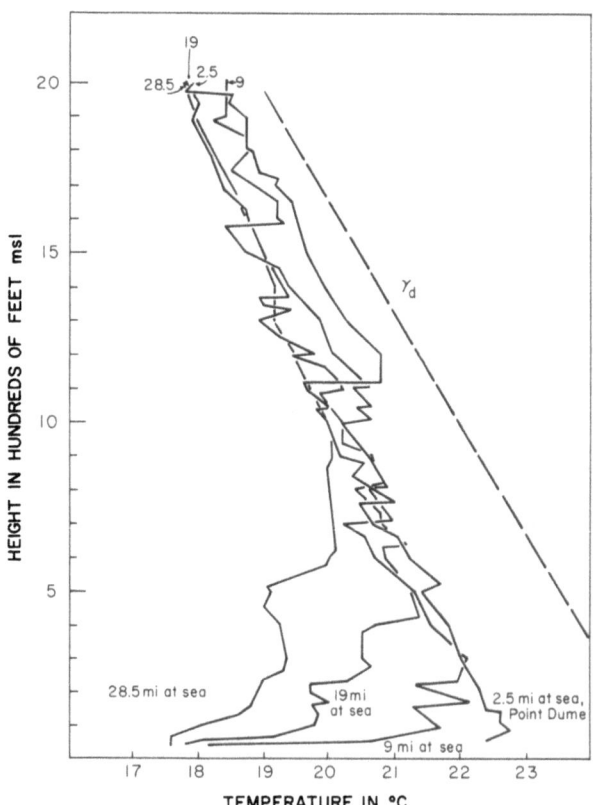

Fig. 6. Temperature soundings on a line from Point Dume to Santa
    Barbara Island, Feb. 4, 1960.

it must at the same time get shallower. The air at the top of
this volume is being heated by compression as it descends but the
air at the surface remains at the same elevations, is unheated.
This sort of deformation of air volumes is called subsidence and
tends to stabilize air. If carried far enough it will produce a
temperature inversion in the layer. Characteristically subsidence
is associated with anticyclones. Air tends to spiral out away from
the centers of these high pressure areas producing a horizontal
divergence of the air which of necessity is accompanied by sinking
of the air aloft over the anticyclone. The presence of an anticyclone over the eastern United States for a protacted period set
the stage for the air pollution disaster at Donora, Pennsylvania,
in 1948. And it is the large semipermanent anticyclone off the
coast of California that provides Los Angeles with its persistent
summer inversion.

Usually inversions are the product of a combination of these
stabilizing processes operating simultaneously. An example frequently observed in California in the winter time is the one which
provides the Great Central Valley with its persistent foggy spells.
If sufficient pollution were available they might be smoggy spells
instead. In this case the "pollutant" is water vapor advected in
from the Pacific behind a storm. It is an anticyclone following
the storm, then stagnating over the western United States, that
produces the subsidence aloft. It is the radiation losses to the
clear skies behind the storm that supplies the strong cooling at
and near the ground, the weak winds below the anticyclone that
account for the lack of mechanical mixing that otherwise would
tend to destroy the inversion. And it is the encircling mountains
that shed their thin mantle of nocturnally chilled air in the form
of drainage currents down the slopes and canyons to add yet more
cold air to the already substantial lake of cold air in the valley.
Chilled air below--warmed air above. Result: a strong inversion
that clamps the lid down on a vast lake of cold air so that it
cannot escape vertically, just as the encircling mountains prevent
its escape horizontally. After a night or two of chilling, the
air is cooled to its dewpoint and fog forms. If the preceding
storm left the landscape drenched the fog forms even earlier by
virtue of the extra water vapor introduced via daytime evaporation.
The fog, thus formed, protects the inversion by day by practically
eliminating heating at the ground, reinforces it by night by radiative
cooling from the top of the fog (base of temperature inversion).
It is a self perpetuating situation. The fog thickens and deepens
as days, sometimes weeks, pass and only the eventual departure of
the anticyclone and the arrival of brisk winds brings an end to a
condition fraught with air pollution potential. In the long run
the fog may turn out to be a mixed blessing, withholding that
essential ingredient of photochemical smog, solar radiation.

By contrast, the persistent summer inversion over coastal California, though formed by very pronounced heating (subsidence) aloft, is not reinforced by cooling at the surface. Instead it is opposed by weak heating there. As the air approaches North America, taking a wide curving path around the northeast side of the North Pacific sub-tropical anticyclone, it moves from colder to warm parts of the ocean. (See Figure 7.) The resultant warming from below produces convective mixing that distributes moisture evenly through the lowest 500 to 1000 ft of the atmosphere, and it establishes a nearly normal lapse rate through this "marine" layer. So the base of the inversion does not appear at the surface but has as its lower limit the top of this shallow marine layer that has come into equilibrium with the ocean's surface. It is this marine layer that, upon arrival over the Los Angeles Basin, becomes transformed into the smog layer. Along only one short part of the air's oceanic trajectory is this weak warming from below reversed and that is along the central and northern California coast where an area of upwelling produces quite cold surface water. The cooling of the marine layer from below along this part of the California coast accounts for the high incidence of fog there.

The inversion forming processes usually have a seasonal variation. For example, during the winter season the Pacific subtropical anticyclone is much weaker and positioned further south with the result that the inversion along the California coast does not occur so frequently nor is it as strong. Occasional winter storms bring lively winds and relatively unstable layers of great depth. With this sort of horizontal and vertical ventilation the smog virtually disappears for Los Angeles. Put more accurately, it is diluted beyond recognition. Ironically, however, the anticyclones following these storms sometimes stagnate over the Great Basin and bring to Los Angeles yet another low, strong inversion. Happily it usually supplies strong winds from the northeast, the famed "Santa Ana", and the associated horizontal ventilation normally prevents the accumulation of high concentrations of smog beneath the inversion but not always. Terrain effects interfere. The San Gabriel Mountains act as a barrier 40 miles long and over a mile high upwind of the city. Air motions in the resultant wind shadow are greatly reduced. In fact, they are sufficiently weak so that a sea-breeze develops in this area during the warmest hours of the day, reversing the direction of the flow, (See Figure 8), often bringing previously polluted air back over the source area for a new load of contamination. It is this weather situation that usually produces Los Angeles' highest smog concentrations.

When the weather situation provides just the opposite of the inversion producing processes: cooling from above and heating from below, inversions, if present, become destroyed. Cooling aloft occurs when the atmosphere is stretched in the vertical, the

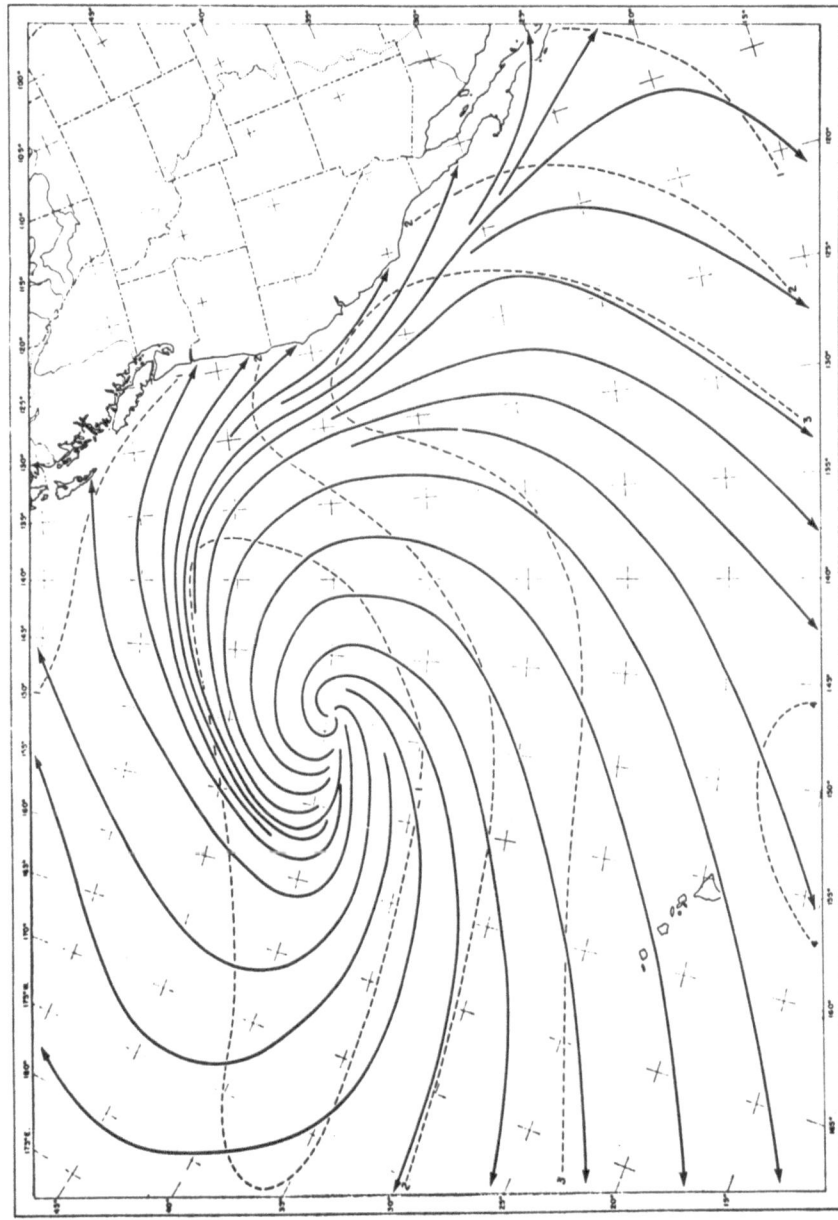

Fig. 7. Normal resultant wind streamlines and isotacks for the eastern Pacific Ocean in July.

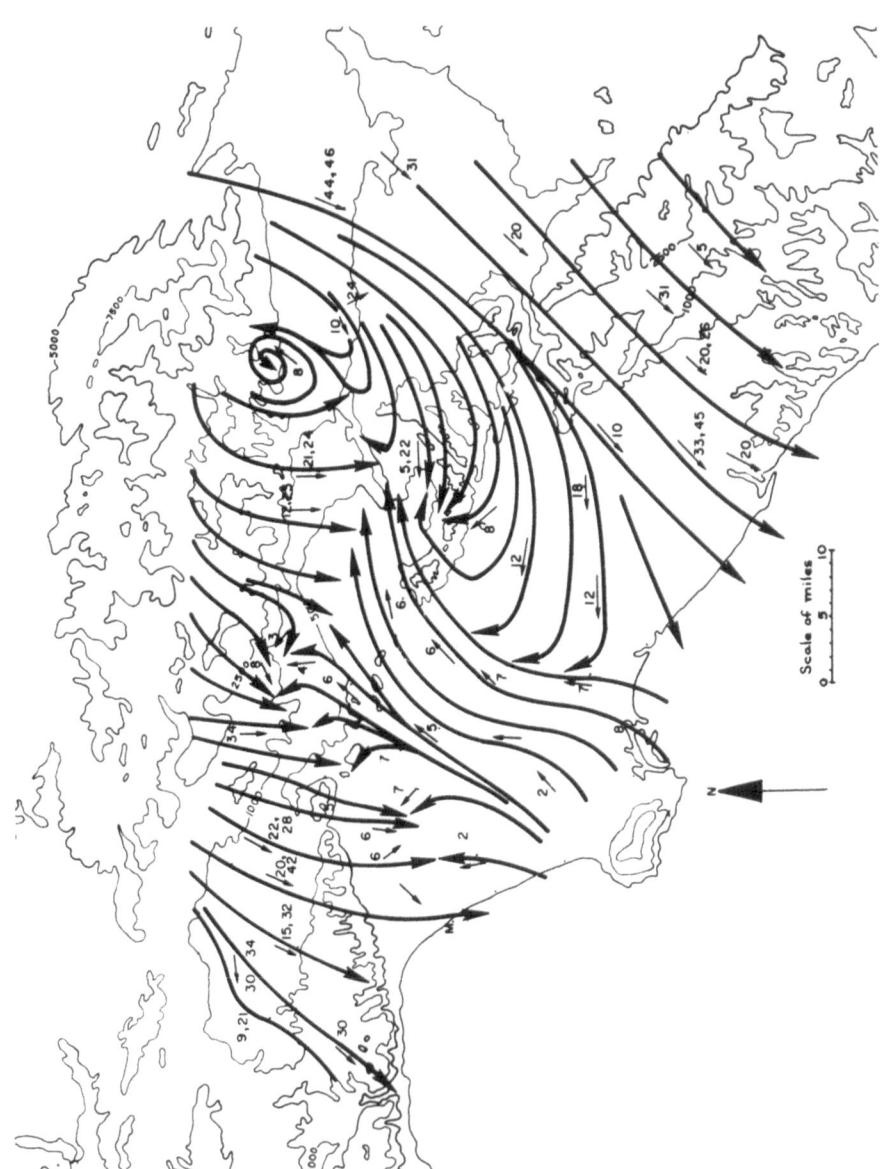

Fig. 8. Surface wind pattern for Los Angeles Basin, 1500 PST, Nov. 5, 1961.

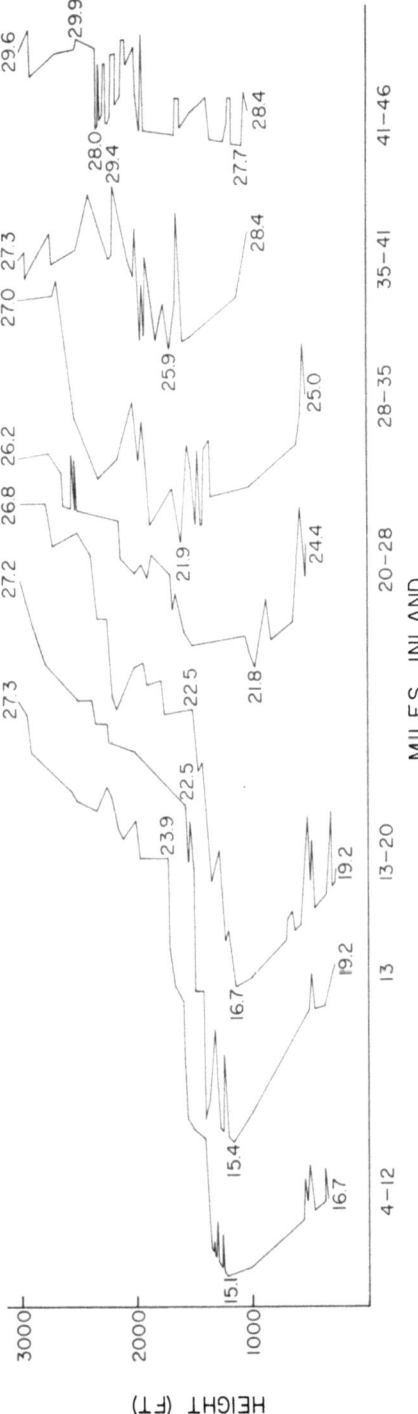

Fig. 9. Temperature soundings along the Santa Clara River Valley in Southern California, 1300–1330 PST, July 28, 1961.

opposite of subsidence. This generally is the case in the vicinity
of cyclonic storms and accounts, in large part, for the comparative
lack of low inversions there. Heating from below is a regular daytime occurrence over land in the absence of clouds and accounts for
the rapid weakening, if not complete destruction, of low inversions
during the day. Figure 9 illustrates this point by showing air-
plane temperature soundings through the marine layer and sub-tropical
inversion at successively greater distances inland on a typical sea-
breeze day in southern California. At the most inland station,
some 40 miles from the coast, where the air has had the longest
history of heating from below, the inversion has almost completely
disappeared. The small excursions to the right in the temperature
vs. height traces in the marine layer denote levels where the plane
penetrated heated convective elements.

Concluding Remarks

    As the weather varies from place to place over the face of the
earth so also does the atmosphere's potential for dispersing air
pollution. By virtue of one city's predisposition for anticyclonic
weather it may suffer more air pollution than another city more
often exposed to cyclonic storms. By the same token, since weather
at any given location varies from hour to hour, day to day, and
season to season, so does the atmosphere's susceptibility to air
pollution vary with time. Certain times of the day and certain
times of the year are definitely more pollution prone than others.
These variations in the atmospheric configuration in concert with
the characteristics of the sources of pollution prescribe the
places and times where the high concentration of pollutants will
occur. The fact that these occurrences show up with a high frequency at an increasing number of locations as the years pass cannot
be interpreted as a new trend in atmospheric behavior. The atmosphere as observed over periods of years, even hundreds of years,
is not changing much. It is the rate at which pollutants are being
introduced into the atmosphere that is changing. Where sources are
absent the air is still pure. For example, the fantastic potential
for contamination of the calm air trapped in the vast radiation inversion layer over the winter Arctic ice fields is not realized
except in a few densely populated spots, such as Fairbanks, Alaska.
It is worth noting at this meeting concerning combustion-generated
air pollution that the combustion product most troublesome to Fairbanks on these extremely cold, stable, windless occasions is water
vapor. At these bitter temperatures, below -40 deg C, it takes
very little water vapor to completely saturate the air. With the
injection of combustion products Fairbanks disappears beneath its
own thick ice fog in the middle of an otherwise clear Arctic expanse.

    In other locations the situation is reversed. Extensive, pro-
lific sources of air contamination have poured forth their aerial
effluent for many decades, (some for centuries), into fairly well

stirred air with no serious pollution effect. In most cases their
rate of release of contaminants has only rarely exceeded the local
atmosphere's rate of dispersal. Each year, however, the former
rate catches and exceeds the latter for a few more cities and on
more frequent occasions. Until we are able (also willing) to control the atmosphere in favor of better dilution or somehow control
the sources of contamination, ever increasing areas of the earth
will spend every increasing amounts of time beneath a polluted pall.

Suggested Reading:

On turbulent diffusion of pollutants:

> Priestley, C. H. B., R. A. McCormick and F. Pasquill, 1958:
> "Turbulent Diffusion in the Atmosphere", Technical Note No.
> 24, World Meteorological Organization, WMO - NO.77.TP.31

On formation of Sub-tropical Inversion:

> Neiburger, M., D. S. Johnson and C. Chien, 1961: "Studies
> of the Structure of the Atmosphere over the Eastern Pacific
> Ocean in Summer, Part I. The Inversion over the Eastern
> North Pacific Ocean", University of California Publications
> in Meteorology, Volume 1, No. 1

On terrain influences and local winds (non-technical):

> Edinger, J. G., 1967: "Watching for the Wind", Anchor
> Books, Doubleday and Company, Garden City, New York

CHEMICAL ANALYSIS OF AIR POLLUTION SOURCES

L. S. Caretto

Professor of Mechanical Engineering

University of California at Berkeley, California

Many techniques are available for analyzing the emissions from industrial and transportation sources. Most of these techniques involve instrumental analysis, as opposed to the traditional "wet chemical" methods. In order to recognize the advantages and limitations of a given analytical method it is important to understand the principles of the method. The first part of these notes discusses chemical instrumentation in general; the second part then describes applications of these techniques to analysis of various emissions.

There are many books available describing many forms of instrumental analysis (1, 2, 3) as well as texts devoted to a single technique. Reference to such texts will expand on the material in these notes.

The references given in the second part for analysis of specific compounds are intended to be selective. There are many journals which will contain articles on chemical analysis of air pollution sources. A good starting point for finding material published in recent years are the review articles by Altshuller (4,5).

PART I   PRINCIPLES OF CHEMICAL INSTRUMENTATION

In the chemical analysis of multicomponent mixtures two basic approaches are available. The mixture can be separated into individual components each of which can be identified and measured quantitatively, or a method of analysis which is sensitive to only one compound (or group of compounds) can be used. In the first method one must be sure that the separation is complete, i.e.,

that what is claimed to be a pure component of the mixture is actually so. In the second one must be sure that the analytical
method is accurate independent of other components in the mixture.

Flame Ionization

Hydrogen flame-ionization detectors (FID) are based on the
observation that hydrogen/air flames produce negligible ionization;
when organic compounds are introduced into the flame, however, a
measurable amount of ions are produced. If a potential is imposed
across the flame the ions will form a current whose value is proportional to the number of ions and hence proportional to the concentration of the organic compound present. The operation of a
hydrogen flame-ionization detector is illustrated in Figures 1 and
2.

Figure 1. Flame-ionization detector.

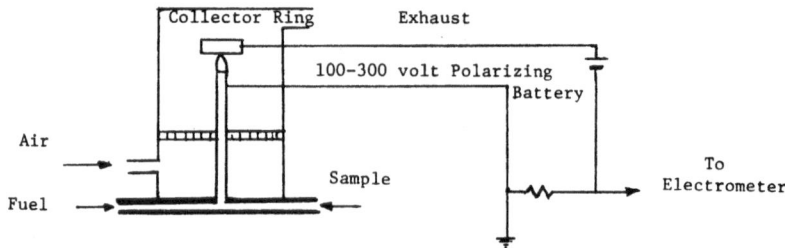

Figure 2. Burner detail.

The sample and hydrogen are mixed and fed to the burner tip. The air flows around the outside of the burner forming a hydrogen diffusion flame. A potential is imposed between the burner tip and an ion collector causing an ion current to flow. This current can be measured by a sensitive electrometer.

The physical process by which the ions are formed is not clearly understood. The flame temperature is too low to account for the amount of ionization by themodynamic equilibrium calculations. Nevertheless, the hydrogen flame detector has been shown to be very sensitive, capable of measuring concentrations as low as 1 part per billion. The hydrogen flame detector was first described by McWilliam and Dewar (6) for use as a gas chromatography detector.

In the operation of the detector it is important that the flow rates of fuel, air, and sample be carefully controlled. Although some slight tolerance is possible in the fuel and air flow rates, the measured concentrations are directly proportional to the sample flow rate, hence this must be most carefully controlled. Some instruments use pure hydrogen as the fuel, others use a mixture of hydrogen and nitrogen.

The sensitivity of the FID to a given hydrocarbon is generally proportional to the number of carbon atoms present in the hydrocarbon. This holds for different classes of non-oxygenated hydrocarbons (paraffins, olefins, actylenes, and aromatics). The method is not so sensitive toward compounds which have oxygen or nitrogen atoms in them, and the relative response of these compounds is not in proportion to the number of carbon atoms in them. Ackman (7) has tried to correlate relative response of hydrocarbons to flame ionization detectors on the basis of the weight percent carbon content of certain "fundamental groups."

Flame ionization analysis of exhaust gases gives a method which is selective for hydrocarbons as CO, CO_2, oxides of nitrogen, water, and nitrogen have no effect. In measuring total hydrocarbon concentrations the instrument is calibrated with a known concentration of a given hydrocarbon (typically n-Hexane) and total hydrocarbon concentrations are reported as equivalent parts per million of the standard hydrocarbon. Jackson (8) has studied the use of flame-ionization detectors for this purpose. Using n-Hexane to calibrate the insturment, he measured the relative responses of various hydrocarbons defined as follows.

$$\text{RELATIVE RESPONSE} = \frac{\text{MEASURED HC CONCENTRATION AS EQUIVALENT n-HEXANE}}{\text{KNOWN HD CONCENTRATION}} \cdot \frac{6}{\text{No. of C atoms in HC}}$$

For example, if a standard sample of 100 ppm of propane (C_3H_8) were measured as 52 ppm equivalent hexane, propane would be said to have a relative response of $(52)(6)/(100)(3) = 1.04$. He found that the relative response of the 15 hydrocarbons he studied was close to 1. (The study included 9 paraffins, 3 olefins, 2 acetylenes, and 2 aromatics.) Generally the relative responses of the hydrocarbons were independent of the flow rates (fuel, air, and sample) used, (i.e., the calibration and measurement, made at the same flow rate, were independent of the levels of flow used, but both measurements must be made at the same set of flow rates.) Some anomalous results were noted for the acetylenic compounds, however. Two different commercial analyzers were used in this study; one instrument showed only a slight dependence of relative response on flow rates. In the other insturment, however, the relative response of acetylenic compounds was found to be strongly dependent on the ratio of hydrogen flow to sample flow. Best results were obtained (i.e., relative responses closest to 1) when the hydrogen flow rates was four times the sample flow rate.

In the same study mixtures of six and eight hydrocarbons were prepared. The expected total concentrations, corrected for relative responses, were 2942 and 2864 parts per million carbon atoms. The measured concentrations were 2970 and 2910 ppm C, respectively. This indicates that the concentrations of individual hydrocarbons are additive.

Jackson also showed that oxygen in the sample caused an interference in the hydrocarbon readings. This interference was found to be a function of the oxygen concentration and the ratio of flow rates: Hydorgen/sample. For a hydrogen/sample flow rate ratio of 4 the average error due to oxygen interference was less that $\pm$ 2% for oxygen concentrations of 5% mole or less. This indicates that exhaust streams that are expected to contain large amounts of oxygen should be calibrated with oxygen-containing calibration gases. This should be especially important for Deisel engine and gas turbines. Pearsall, (9) however, has described an FID for use with diesel exhaust which is independent of the oxygen content of the sample.

Commercial instruments for analysis of total hydrocarbon concentrations in automobile exhaust are thermostated so that they can operate at elevated temperatures (ca. 150°C). In this way it is not necessary to condense out the water before feeding the stream to the analyzer. It is necessary, however, to carefully filter the exhaust stream to avoid clogging the burner. Since the burner tip forms part of the electrical circuit which performs the measurement, it is important that this tip be kept carefully clean. (Measurements are even sensitive to finger-prints left on the burner tip.)

The flame ionization detector is a very sensitive detector and can be used to detect very low (ca. 1 ppb) concentration of hydrocarbons. When being used to measure very low concentrations, several precautions must be taken:

The flow rates must be carefully regulated to avoid disturbances which could alter the measured concentration.

The hydrogen and air used must be ultra pure to avoid spurious signals from contaminants in these gases. (Some manufacturers produce pure grades of gases especially for these detectors.)

Lines used for the various flowing streams must be scrupulously clean.

These considerations are espeically important when the flame ionization technique is used as a detector in gas chromatography (see below).

In summary, flame ionization offers a rapid and accurate method for measuring the total hydrocarbon concentrations in exhaust streams from combustion devices. Care must be taken to assure that all operating parameters of the instrument are the same for both calibration and measurement of exhaust. Special precautions should be taken if it is known that the exhaust stream contains an unusually large amount of oxygen.

Gas Chromatography (10,11)

Of all the various methods discussed here gas chromatography is perhaps the best method from the standpoint of accuracy, versatility, and definite qualitative identification of the compounds. Rather than giving a measure of a specific compound in a mixture, gas chromatography provides an actual separation of the components of the mixture. Each component can be then identified and quantitatively measured. The disadvantages of chromatographic techniques is that they are restricted to batch sampling procedures and can be time consuming.

Basic Theory

The operation of a gas chromatograph is centered on the separation column. The operation of this column is based on simple rules of thermodynamic phase equilibrium. The column is packed with a porous material which may be a solid or a solid impregnated with a liquid. The former are called adsorption columns; the latter, partition columns.

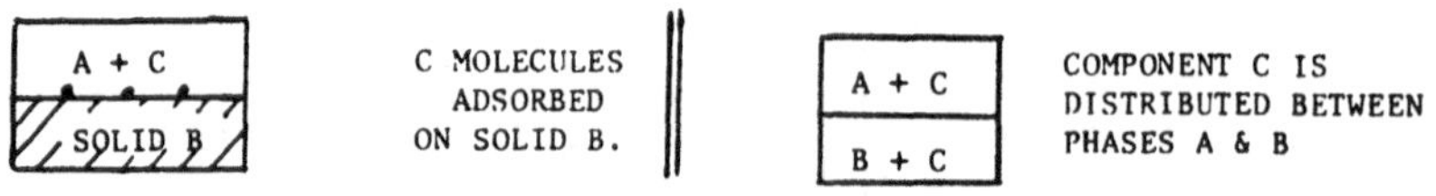

A + C ///SOLID B///	C MOLECULES ADSORBED ON SOLID B.	A + C / B + C	COMPONENT C IS DISTRIBUTED BETWEEN PHASES A & B

Figure 3. Adsorption and partition.

The diagrams above illustrate the difference. In adsorption, equilibrium obtains between a component C which is present in the fluid phase A and adsorbed on the solid B. At low concentrations of C in the fluid phase the equilibrium can be represented by Henry's law

$$\Theta_C \;=\; K \; c_C^{(A)}$$

where Θ_C represents the amount of C adsorbed on the solid B expressed as a surface coverage, $c_C^{(A)}$ represents the concentration of C in the fluid phase and K is a constant which is a function of temperature and pressure. (The dependence of K on the total pressure is very slight; it should be noted that K does not depend on composition.)

In partition equilibrium Nernst's law applies:

$$c_C^{(B)} \;=\; N \; c_C^{(A)}$$

Here N, the so-called partition coefficient, is a function of temperature and (slightly) of pressure. The partition that is carried out in a chromatographic column is probably different from that carried out in an ordinary separatory funnel as the surface to volume ratio is much greater in the chromatographic column.

A schematic diagram of a simple gas chromatographic system is shown below.

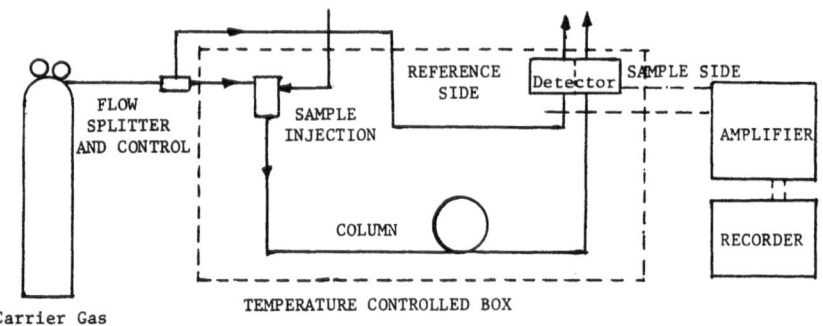

Figure 4. Typical arrangement of a gas chromatograph.

In the actual operation a carrier gas stream is constantly flowing through the column. (Helium is typically used as a carrier gas but other gases can be used. This will be discussed later.) The sample to be analyzed is injected into the carrier gas just before the column. The components which have little affinity for the column material pass swiftly through the column; those which tend to stay on the column packing will pass more slowly through the column. By a judicious choice of column material and operating temperature, a complete separation of the components can be made.

The detector at the exit of the column is set to measure differences between the carrier gas and the components of the sample gas. As the separated components pass through the detector they produce a signal which is recorded as a function of time. A typical chromatogram (i.e., recording of the detector output) is shown in Figure 5.

The time difference between the injection of the sample and the appearance of the peak is called the retention time of the peak. In many samples the first peak represents air in the sample which passes through the column with minimal contact. This gives a measure of the residence time, i.e., the time required to flow through the column void space independent of any adsorption on the column material. Retention times are sometimes measured as the time between the appearance of the air peak and the appearance of a given peak.[*] Since

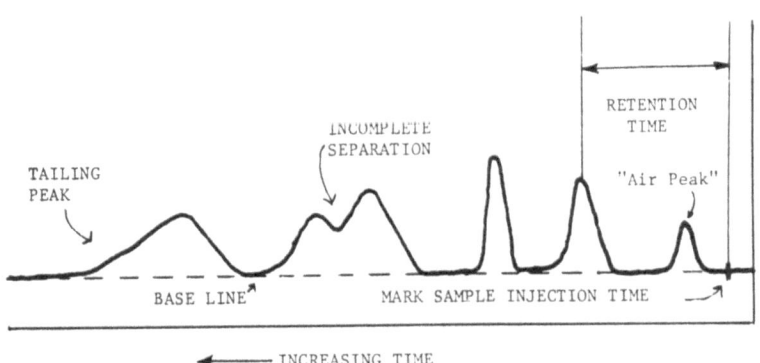

Figure 5. Sample chromatogram.

_____
[*]The term "Air Peak" is laboratory jargon for the more formal name "unadsorbed Gas Peak."

the peak represents a time record of the concentration of the component in the carrier gas as that component passes through the detector the amount of that component present in the original sample is proportional to the area under the peak.

Qualitative identification is made by comparing the retention times of the peaks in the unknown sample to those of peaks from a previous chromatogram with known compounds. Thus the use of the gas chromatograph implies a knowledge of the possible components present in the stream since a calibration is necessary before qualitative identification can be made. To verify a calibration curve the separated components from the unknown can be trapped and analyzed by some other method (e.g., infrared analysis or mass spectrometry) to verify their identification. This precludes the possibility that some unknown compound in the sample will have the same retention time as one of the calibrated compounds and result in a false identification. When two compounds have the same retention time there are several alternatives available:

1. If the identification of each compound is not important the results can be reported as the concentration of both compounds.

2. A different column may be able to effect the separation not obtained with the first column.

3. One of the compounds can be selectively removed from the sample stream. (This is especially useful as a means for removing certain types of hydrocarbons; e.g., some trapping columns will remove olefins and allow paraffins to pass.)

Often a single column in a isothermal environment is not sufficient to affect the desired separation. If a column which as a strong affinity for the first few peaks is used, the later peaks will require a prohibitively long time to elute and will exhibit severe tailing, giving poor separation and making quantitative analysis difficult. On the other hand, if the column has a very low affinity for the first few components to elute, they will not be properly separated and a good analysis will only be possible for the later peaks. These problems can be avoided by using multiple column systems and temperature programming.

In the latter method the column is heated at a carefully controlled ("programmed") rate. This means that the first peaks to elute traverse the column at a low temperature, such that they will tend to remain on the column and give a good separation. As the temperature is increased the components have less affinity for the column material and tend to pass through faster. This permits the heavier components of the unknown mixture to be eluted in a reasonable time and form symmetrical peaks.

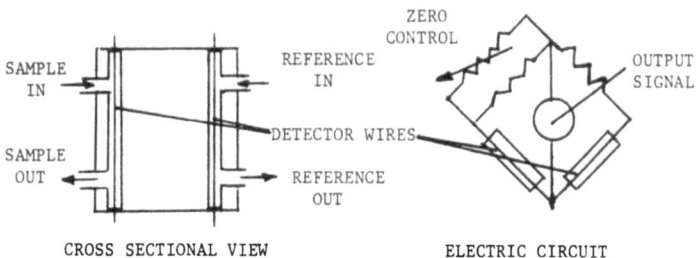

Figure 6. Thermal conductivity detector.

Detectors

The simplest detector used on gas chromatographs is the thermal conductivity detector. A schematic of a basic thermal conductivity detector is shown in Figure 6. In this detector, which is usually operated as a differential detector, an electrically heated wire is mounted in the path of a flowing gas. The temperature, and hence the resistance, of the wire is a function of the physical properties (mainly the thermal conductivity) of the gas flowing over it. Two such wires are used in the detector; carrier gas flows over one wire and the exit stream of the chromatographic column flows over the other. The two wires are arranged to form a Wheatstone bridge. When the gas flowing over both wires is the same there is no difference in the resistance and the bridge is balanced. When a component is eluted from the column the difference in thermal conductivity causes a difference in temperature which is converted into an electronic signal due to the change in the resistance of the wire.

The thermal conductivity detector is almost a universal detector. Its ability to distinguish compounds is related to the difference in thermal conductivity between the carrier gas and the compound to be detected. For this reason helium is generally used as the carrier gas when thermal conductivity detectors are employed. (Helium has the highest thermal conductivity of all gases except for hydrogen; the latter is generally not used for safety reasons. If hydrogen is to be analyzed by gas chromatography using a thermal conductivity detector, it is advisable to employ some other carrier gas since mixtures of hydrogen and helium have anomalous values of thermal conductivity.)

The flame-ionization detector mentioned previously is an accurate, sensitive detector for the chromatographic analysis of hydrocarbon mixtures. Its sensitivity requires great precautions to assure that the chromatographic system is scrupulously clean.

In addition, it has been noted that some partition columns tend to
bleed off the packing material, giving spurious signals. Thus flame
ionization detectors are best used with adsorption columns. If a
partition column must be used, it is usually best to bake it out for
several hours before actual use. It has been found that the FID
is sensitive to the thermal conductivity of the carrier gas (12)
and it has been shown that carrier gases with lower thermal conductivities give best results. For separation of hydrocarbon mixtures on alumina columns nitrogen has been recommended as the best
compromise carrier gas (13). Flame ionization detectors are about
1000 times as sensitive as thermal conductivity detectors.

Besides these two detectors many other types are available.
The thermal conductivity and flame-ionization are by far the most
popular. Information about other types of detectors can be found
in manufacturer's literature. A discussion of the various types of
ionization detectors has been given by Lovelock (14).

Sample Injection

To have reproduceable analyses the amount of sample introduced
each time must be reproduceable. Samples are usually introduced by
means of a hypodermic syringe or a gas sampling valve. For gaseous
samples the sampling valve is usually the most effective way of
entering an accurate sample.

Figure 7 shows the basic idea of a gas sampling valve. The
sampling valve usually consists of two loops; while the carrier
gas is flowing through one loop to the column the other loop can be
filled with sample. By rapidly switching the valve the new sample
can be introduced into the column.

When hypodermic syringes are used to introduce the sample the
carrier gas flow goes through an injection port fitted with a rubber septum. (See Figure 8.) The sample is injected into the carrier gas stream at this point. Typical sample sizes are fractions
of milliters for gases and microliters or fractions of microliters
for liquid samples. For liquid samples it is importnat that the

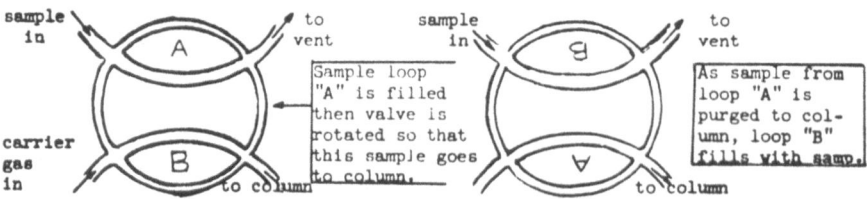

Figure 7. Gas sampling valve.

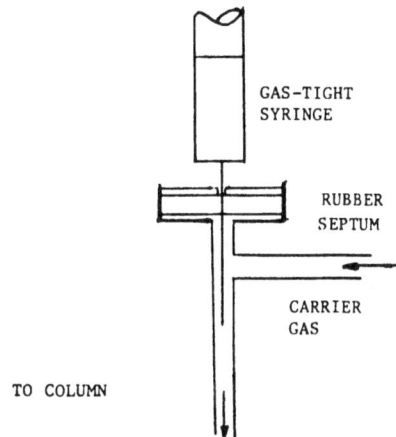

GAS-TIGHT
SYRINGE

RUBBER
SEPTUM

CARRIER
GAS

TO COLUMN

Figure 8.   Syringe injection.

injection port be heated to give rapid vaporization of the liquid.

Carrier Gas

The carrier gas is determined by the type of detector to be used. Helium is an almost universal carrier gas, but nitrogen is typically a better carrier gas for flame ionization detectors. In any case, the carrier gas to be used must have a purity commensurate with accuracy of the analysis to be made. An important factor in obtaining reproduceable elution times is the reproducibility of the carrier gas flow rate. Most instruments are carefully constructed to assure repeatable flow rate settings.

Columns

The various types of columns that have been used forms an extensive list. When trying to decide on a column for a given separation the best references are manufacturer's literature for the literature describing a previous similar analysis. As mentioned above, the two general types of columns are partition columns and adsorption columns. These have been generally prepared by packing tubing (copper, plastic, or stainless steel) with fine particles of packing material. More recently, capillary columns, formed by coating the interiors of narrow capillaries have been shown to give very efficient separations. Such columns have been shown to give excellent separations of complex hydrocarbon mixtures (15).

Conclusions

Gas chromatography offers a versatile tool for the analysis of

exhaust streams. It is perhaps the most effective method available
for definitive identification of all components, especially unburned hydro-carbons, present in these streams. Used in conjunction with infra-red analysis or a mass spectrometer, it provides a
powerful research tool for investigation of the chemistry of the
combustion process. Given the myriad of compounds present in combustion processes, the gas chromatograph can provide the most nearly
complete analysis available.

Spectroscopic Analyses (16)

The operation of various photometric and spectroscopic techniques is based on the observation that the spectrum of light absorbed as a function of frequency is different for each compound.
This provides a means of qualitative identification of a compound.
The amount of light absorbed by a sample cell containing a given
compound is related to the concentration of that compound, allowing
quantitative analyses to be made.

These analyses can be divided into two broad classes: dispersive where only a narrow region of the spectrum is used to analyze
the compound and non-dispersive where a wide frequency range is
used. In the former, the light from a source is said to be "dispersed" into various frequencies. Dispersive analyses find more
use as a laboratory tool, while non-dispersive analyses have long
been a useful tool in air pollution analysis.

Dispersive Analysis

The theory of the occurrence of spectra is based on quantum
transitions between various energy levels in a molecule. Radiation
will be absorbed or emitted only at frequencies that are related
to the change in energy levels. For complicated molecules there
are overtones and coupled energy transitions (e.g., rotational-
vibrational)such that the final spectrum forms a distinctive pattern.

The basic features of a dispersive spectrometer are illustrated
in Figure 9.

The basic sections are the source, which provides the radiation,
the monochrometer, which disperses the radiation into narrow bands,
and the detector, which measures the radiation transmitted. An
arrangement of mirrors and slits is used to focus the radiation and
provide a constant amount of radiation over the range of the instruments. Mirrors are used rather than lens, since they have less
aberration. Either a prism or a diffraction grating is used to
provide a narrow frequency band. The diffraction grating will
produce a constant band with over the entire frequency range. The
band width of a prism monochrometer varies strongly with the fre-

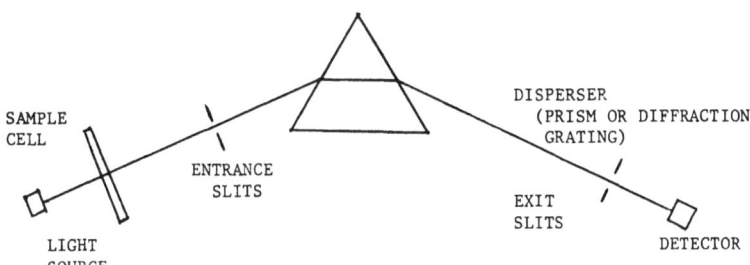

Figure 9.   Schematic of spectrometer.

quency.

    The electromagnetic spectrum is shown below.

WAVE-LENGTHS (λ)					
10^{12}	10^8	10^4	1	Å	(Angstrom unit)
10^{11}	10^7	10^3	10^{-1}	$m\mu$	(milli-microns)
10^8	10^4	1	10^{-4}	μ	(microns)
10^4	1	10^{-4}	10^{-8}	cm.	

RF	MICRO WAVE	IR	V I S	UV		RAYS

10^4	10^8	10^{12}	10^{16}	10^{20}

FREQUENCY   (CYCLES/SECOND)   (ν)

The longer wavelengths (lower frequencies) are radio waves (RF for Radio Frequency) and microwaves. At increasing frequencies (de-creasing wavelengths) the spectrum contains the infrared (IR), visible (VIS), and ultraviolet (UV) regions. γ-Rays occupy the very high frequency region of the spectrum. There are many units used for wavelength. The units used in a given discussion are ty-pically those which give numbers greater than unity.

$$1 \text{ cm} = 10^{-2} \text{ meters}$$

$$1 \ \mu \ (\text{micron}) = 10^{-6} \text{ meters}$$

$$1 \ m\mu \ (\text{millimicron}) = 10^{-9} \text{ meters}$$

1 Å (Angstrom unit) = 10^{-10} meters

The frequency, ν , and wavelength, λ , are related by the simple relation $\nu = c/\lambda$, where c is the speed of light. Another commonly used measure is the reciprocal wavelength $1/\lambda$, usually reported in cm^{-1}. This is readily converted to a frequency by multiplying by c.

The frequency ranges usually used in chemical analysis include the ultraviolet, visible, near infrared and far infrared. The largest portion of spectroscopic analysis is done in the near infrared, in particular in the region between 2.5 and 15 microns. A sample spectrum obtained by Sawyer (17) is shown in Figure 10. The absissia is the reciprocal wavelength or "frequency" in cm^{-1}. The region covered, from 4000 cm^{-1} to 650 cm^{-1} is equivalent to a frequency range of 12 x 10^{13} cycles/sec. to 19.5 x 10^{12} cycles/second. It can also be expressed as a wavelength range of 2.5 to 15.4 microns. The ordinate on these spectra is a measure of the light transmitted. If the power impinging on the sample is denoted by P_O and the power leaving the sample is denoted by P, the transmittance or fraction of power transmitted is defined to be

$$T = P/P_O$$

The absorbance is defined as follows

$$A = - \log_{10}T$$

Both these units are commonly used. The absorbance is especially used in obtaining spectra of solutions where Beer's Law applies. In this case

$$A = - \log_{10}T = \log_{10}(P_O/P) = \varepsilon sc$$

where s is the path length through the sample cell, c is the concentration of the specie absorbing radiation and ε is a constant independent of concentration and path length. Beer's law is generally applicable only in dilute solutions.

To relate concentration to absorbance or transmittance in a general case, it is usually necessary to first obtain a calibration curve. This is usually done by noting the change in absorbance with concentration of a convenient peak on the spectrum. It is desirable to have a strong peak, i.e., one which will show great change with changes in concentration, and a peak which has minimum overlap with peaks of other species present in the sample.

In order to obtain a quantitative analysis of a multicomponent mixture it is first necessary to have some idea of the compounds

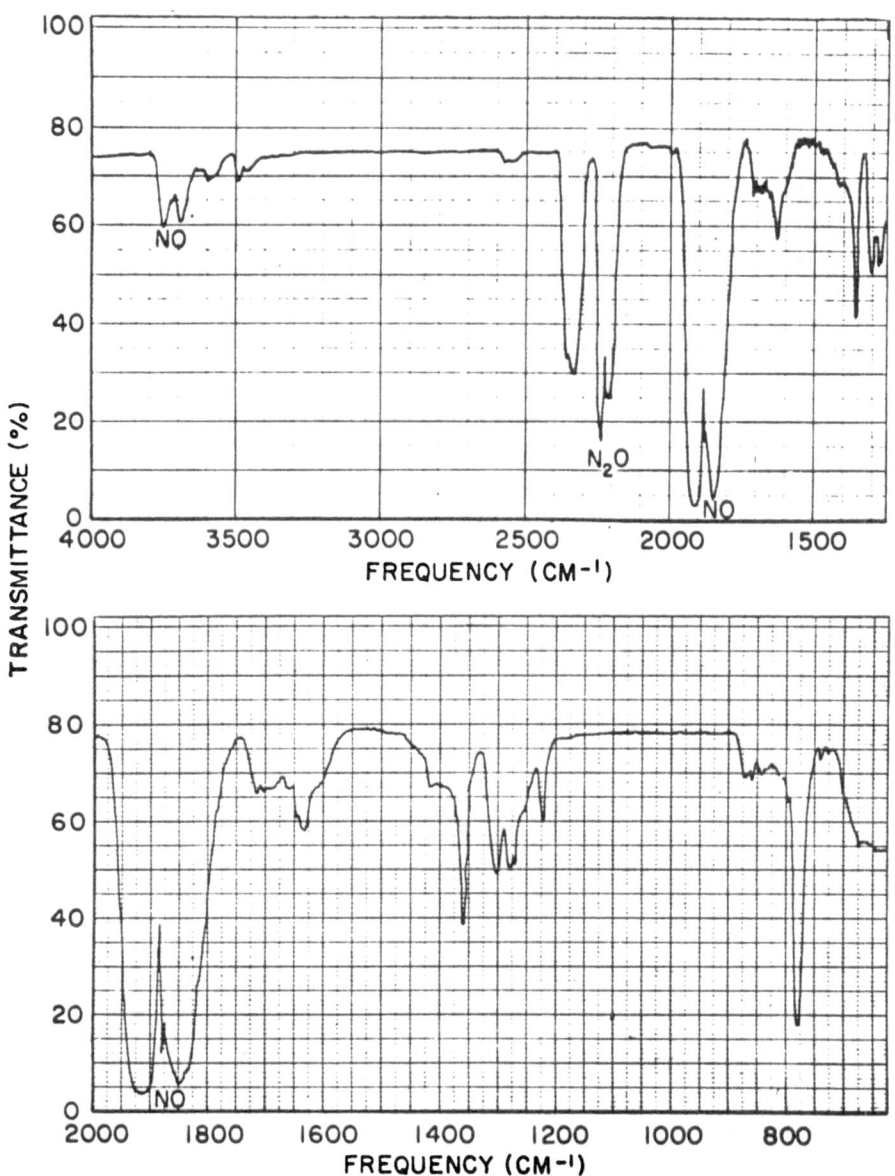

Figure 10.   Infrared spectrum, nitric oxide.

that might be present, and what their spectra are. In analyzing
an n component mixture, it is necessary to measure the absorbance
of n peaks. The peaks are chosen so that there is minimum overlap.
between the peaks of various components in the mixture, but since
there is usually some overlap it is necessary to solve a system of
n algebraic equations to find the concentrations. In this way one
can account for the contribution of all species present to a given
peak.

It is also possible to look at a single portion of the spectrum
to isolate the behavior of one compound if that compound is the
only component of the mixture which absorbs in some spectral region.
With fast response detectors currently available this procedure provides an excellent analysis of the transient behavior of a single
species.

Non-Dispersive Analysis

The non-dispersive infra-red analyzer (NDIR) is perhaps the
most familiar tool used in source analysis. It is, of course, the
legal standard currently in use for measurement of exhaust emissions.
In this method the gas which is being analyzed is used to detect
itself. This operation is shown in Figure 11.

The gas to be measured is used to fill both sides of the detector. These sides are separated by a flexible diaphragm. The
radiant energy striking one side of the detector passes through a
reference cell containing none of the gas to be analyzed; the radiant energy striking the other side passes through a cell through
which a sample stream is continuously flowing. A chopper is placed
in front of the sources so that the radiant energy which arrives at
each side of the detector is the same, the diaphragm will undergo
regular fluctuations at the same frequency that the radiation is
passed by the chopper. The fluctuations of the diaphragm cause the
capacitance between itself and the fixed probe to change. This sets

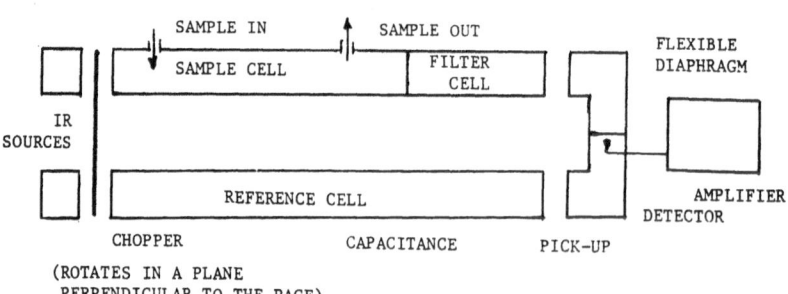

Figure 11. Non-dispersive infrared analyzer.

up a signal which is amplified and recorded. When the sample contains some of the gas that is in the detector some of the radiant
energy passing through the sample cell will be absorbed and the
energy striking the sample side of the detector will be less. This
will cause a difference in the signal detected by the capitance
pick-up and will give a measure of the concentration of the component passing through the sample cell.

An obvious problem with such a detector is that there might
be other components in the sample stream which will absorb radiation
at the same frequencies that the gas in the detector will absorb.
Then this is the case it is sometimes possible to use a filter cell
containing a large concentration of the interfering gas. The analyzer zero is then set with this large concentration of interfering
gas and any interferences in the sample will be small perturbations
on this already large interference and will not be noticed by the
detector. The presence of compounds in the sample which absorb at
other wavelengths is not important since this energy will not be
absorbed by the gas in the detector.

Such detectors are particularly useful for simple gases which
have clear peaks that are not overlapped by peaks of the other components in the sample. In addition, it is not necessary to use
the complete spectrum of the gas in the detector as filters can be
used to block out parts of the spectrum of radiation reaching the
detector. These detectors have proven very useful for the analysis
of CO and CO_2.

When n-Hexane is used in the detector the instrument can be
used to give a measure of the total hydrocarbon concentration as
equivalent n-Hexane. Jackson (8) has measured the relative response
of various hydrocarbons using the non-dispersive infrared in the
same manner as used in studying the flame-ionization detector (see
above). He found that the relative response of the paraffins studied
(except for methane) was close to 1. The response of olefins, acetylenes and aromatics was very low, however, (e.g., acetylene had a
relative response of 0.02). He found that the possible interference
from carbon dioxide could be eliminated by use of a filter cell, but
the interference from water vapor could not be eliminated in this
way. The water vapor content of the stream must be measured and
corrected for by a calibration curve. Jackson also found that the
hydrocarbon analysis, considering relative responses, was additive.

It is also possible to make non-dispersive analyses in the ultraviolet. In this region the detector of the type described above is
not necessary. Instead one can restrict the frequency range of
the radiation by proper choice of light source, filters and photo-
tube detector. Otherwise, all the operating principles described
above can remain the same. Indeed, it is not even necessary to have

the reference cell. The reference cell and signal chopper are used
to give an ac signal to the amplifier since ac amplifiers have better drift free characteristics.

The dispersive instruments provide a versatile tool which is
capable of making chemical analyses without taking a physical sample. Its use for complete analyses of multicomponent mixtures is
somewhat difficult, but it can, under certain conditions, provide
rapid continuous analysis of one component in a mixture.

The non-dispersive analyzers are convenient instruments for
the analysis of a single species, but care must be taken to avoid
interference from other components in the mixture.

MASS SPECTROMETRY (18)

Within recent years mass spectrometry has changed from an ele-
gant research topic to a useful analytical tool. The operation of
mass spectrometers is based on the behavior of charged particles
in electromagnetic fields. The two basic types of mass spectrometers
are the conventional magnetic deflection type and the time-of-flight.
In both cases the sample to be analyzed is initially bombarded with
high energy electrons to form ions, with charge, e. These ions are
then accelerated through a voltage, V, and in doing so they acquire
an energy, eV. This is represented as kinetic energy of the ion,
i.e.,

$$eV = \tfrac{1}{2} mv^2$$

Thus the velocity, v, of the particle of mass, m, at the end of
the acceleration step is given by

$$v = (2V \tfrac{e}{m})^{\tfrac{1}{2}}$$

The velocities of the ions are determined by their ratio of mass-
to-charge, m/e, for a fixed voltage.

In the conventional magnetic deflection spectrometer the ions
next enter a magnetic field of strength, H, where they undergo a
force given by

$$F = eHv$$

The force is perpendicular to both the direction of the magnetic
field and the velocity of the ions causing the trajectories of the
ions to bend. (See Figure 12) The radius of curvature of the ion
trajectory is given by a force balance between the force of the
magnetic field and the centrifugal force mv^2/r, where r is the radius
of curvature. This force balance gives

$$eHv = \frac{mv^2}{r}$$

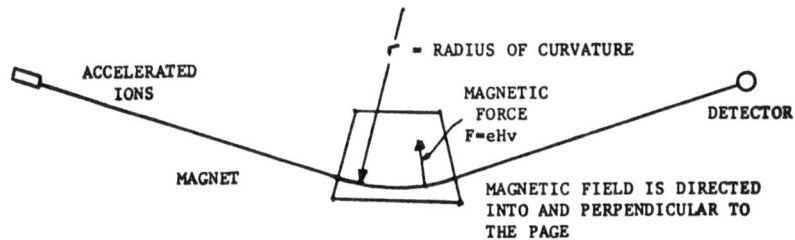

Figure 12. Operation of magnetic deflection mass spectrometer.

and the radius of curvature is then given by

$$r = \frac{mv}{eH} = \frac{m}{eH} (2V \frac{e}{m})^{\frac{1}{2}} = (\frac{2Vm}{eH^2})^{\frac{1}{2}}$$

Thus, for a given value of accelerating voltage and magnetic field strength the radius of curvature and hence the trajectory of the ion is determined by the mass-to-charge ratio of the ion. If one wanted to analyze for a particular m/e ratio it would only be necessary to pick appropriate values of voltage, V, and magnetic field, H, such that the ion trajectory would lead the ion to the detector. Generally, however, a complete mass spectrum is desired. In this case there are two alternatives: the voltage is varied keeping the magnetic field strength constant or the strength of the magnetic field is varied while the accelerating voltage is held constant. If the output of the detector is then recorded as one of these quantities is varied, a typical mass spectrum, such as the ones shown in Figures 13 and 14, is produced.

The height of the peak produced is proportional to the amount of ions with that given mass-to-charge ratio. In identifying the peaks, mass is measured in atomic mass units (a.m.u.) defined by assigning the normal isotope of Carbon, C^{12} a mass of 12 a.m.u. The unit of charge used is the electronic charge. Thus, an ion formed by the loss of a single electron would be said to have a charge of 1; an ion formed by the loss of two electrons, 2; etc. As an example an ion formed by the loss from a methane molecule ($C^{12}H_4^1$; mass = 16 a.m.u.) would have a mass-to-charge ratio of 16.

The simple description given above must be augmented by the fact that during the preliminary ionization step the molecules of the sample tend to decompoze due to the high energy of the ionizing electrons. In addition, some molecules and fragments are doubly (or higher) ionized. Thus, a pure specie would not show a single

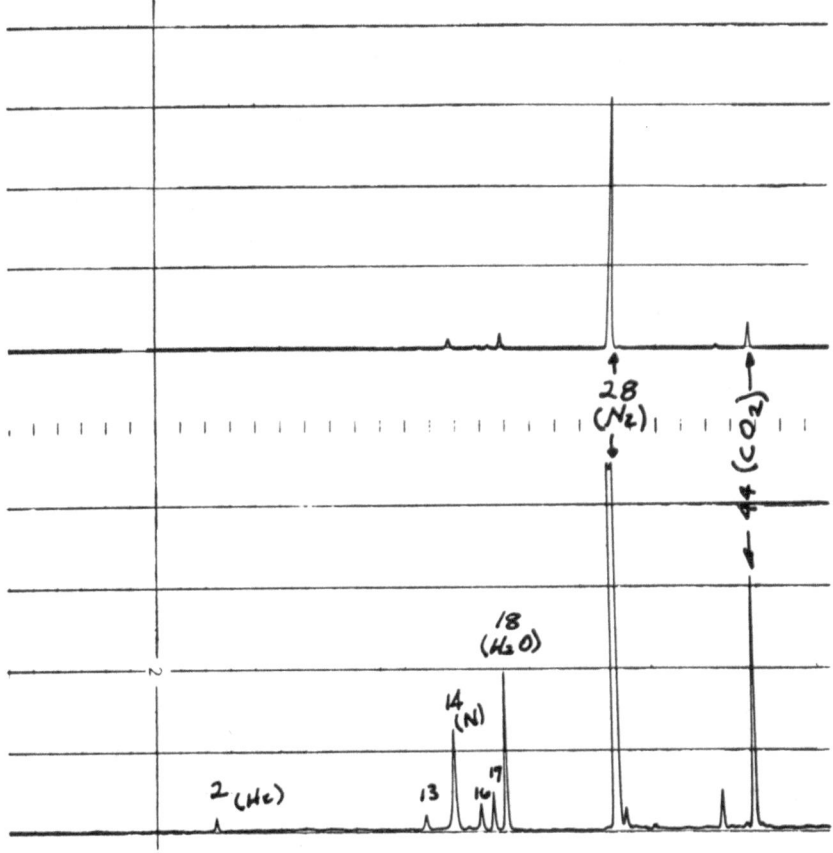

Figure 13. Mass spectrum of propane/air combustion products. Same
spectrum recorded on two channels. Amplification on lower channels is ten times that on upper channel. m/e ratios and species
giving major contribution to some peaks noted on chart.

peak but would show a typical fragmentation pattern. Just as the
infrared spectrum of a molecule serves as a qualitative identification of that molecule, the fragmentation pattern of a pure substance
in a mass spectrograph gives a unique pattern which can be used to
qualitatively identify that specie.

In the time-of-flight mass spectrometer the resolution in space
formed in the magnetic deflection device is replaced by a resolution
in time. The velocity of the ions leaving the accelerating grids
is still given by $v = (2V e/m)^{\frac{1}{2}}$. The ions leaving the accelerating

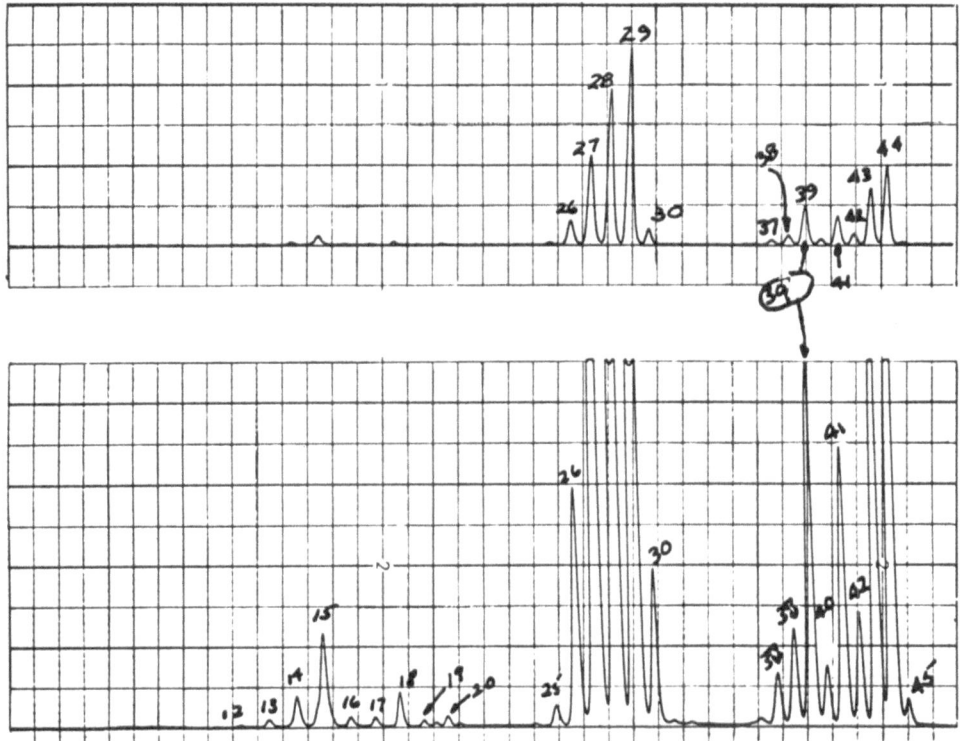

Figure 14. Fragmentation pattern for propane (C_3H_8). Same mass
spectrum is displayed on both channels. Amplification on lower
channel is ten times that on upper channel. Peak m/e ratios
noted on chart.

grids then traverse a drift tube. The time required for a given
ion to arrive at the detector at the end of this drift tube then
depends on the square root of its mass-to-charge ratio. The advan-
tage of the time-of-flight instrument is that an entire spectrum
can be resolved in a time of about 10 microseconds.

In evaluating the performance of both types of mass spectrometers there are three basic factors to consider. The range defines
the span of mass-to-charge ratios that can be detected by the instrument. This is usually given as a mass range (e.g., 1-100 a.m.u. or
1-4000 a.m.u.) since if any singly ionized particle at the upper
limit of the mass range can be detected, the higher ions of this mass
will also be detected. The resolution power of a mass spectrometer
is its ability to distinguish between adjacent peaks. This is ex-

pressed as the difference in mass number between adjacent peaks which can be resolved adn the mass number of the peak (e.g., 1 part in 600). The _sensitivity_ is a measure of the instrument's ability to measure concentration. The mass spectrometer can work with very small samples and is able to detect concentrations of the order of 1 ppm even with these small samples. In the operation of the mass spectrometer there is a constant trade-off of these factors and most modern instruments are adjustable so that the performance can be optimized for the given analysis.

Time-of-flight spectrometers, when used with their high repetition rate do not have the sensitivities of magnetic deflection instruments. When this high rate of reproducing spectra is not important, however, high sensitivities can be obtained from time-of-flight instruments by averaging a large number of spectra with an electronic analog detector.

Analysis of multicomponent mixtures is complicated by the fact that each component of the mixture has its own fragmentation pattern. Thus a given peak in a spectrum from a multi-component mixture may correspond to fragments from several of the components in the sample. In order to use the instrument then, some previous knowledge of the possible components must be known. From published or previously determined fragmentation patterns on pure species the data taken from the peak heights can be reduced by solving simultaneous algebraic equations.

The mass spectrometer can also be set to follow the height of a single peak as a function of time. This is useful when the given peak has significant contributions from only one compound. In this case the height of the peak gives a measure of the concentration of that compound. In using this technique it is necessary to determine whether or not any other compounds will contribute to the peak being monitored (and thus reduce the accuracy of the measurement or require some correction.) For example in a study of nitric oxide concentrations in automobile exhaust (19) the $m/e = 30$ peak, produced by NO^+ ions formed from the normal isotopes ^{14}N and ^{16}O, was monitored. It was necessary to consider interferences from the following, singly-charged ions which also have a mass to charge ratio of 30: Carbon Monoxide $^{12}C^{18}O$, Nitrogen $^{15}N^{15}N$, Ethane $^{12}C_2^1H_6$, and Formaldehyde $^1H_2$ $^{12}C^{16}O$. There could also be interferences from fragments caused by splitting of higher mass molecules.

The combination of gas chromatograph and mass spectrometer provides a powerful analytical tool which could possibly separate and identify almost all the components in an exhaust gas stream. As mentioned previously, the analysis of multicomponent mixtures requires the solution of simultaneous algebraic equations, one for each component in the sample. Even with machine computation this

can become difficult and the spectra produced by a multiple of hydrocarbons could be difficult to interpret. If the gas chromatograph is first used to separate some components of the exhaust and
the effluent of the chromatograph is used as the sample for the
mass spectrometer, the number of components in the sample fed to
the spectrometer would be small allowing for a better and easier
analysis. In addition, this would allow identification of components which could not be separated by gas chromatography.

The problems of interfacing the gas chromatograph and the mass
spectrometer are discussed in the review article by McFadden (20).
There are two basic methods: collection of separated components
from the chromatograph for later introduction to the mass spectrometer and continuous flow of the chromatograph effluent, through a
selective membrane which removes most of the carrier gas, into the
mass spectrometer. The former method provides a better sample; the
latter, being continuous, is easier to operate. Buson and Turner
(21) have discussed the design of a simple trap; Black Flath and
Teranishi (22) have given a discussion of membrane separator interfaces.

The mass spectrometer represents a very powerful analytical
tool. Its use in combustion work has been largely restricted to
flame studies (3) but it can be applied to exhaust measurements.

Other Methods

The use of colormetric methods requires the development of a
new reagent for each application. In these methods the component
to be analyzed is reacted with a solution, specific to that compound, to produce a color. The intensity of the color is a measure
of the species concentration. For the pollutant levels usually
associated with source emissions it is necessary to be careful to
assure that all the compound to be analyzed is reaced with the solution. In addition there is always the possibility of interferences;
i.e. other compounds can react providing the same color and giving
a false measure of the concentration. Many of these methods are used
for analysis of atmospheric concentration levels (i.e. about 1 ppm)
but can be extended to source analysis of higher concentration levels
by dilution of the sample prior to analysis.

Electrochemical methods of analysis are based on measuring
the current generated by cell reactions. For example, if a potential
is applied to two electrodes the normal reaction can be reversed.
This phenomenon is known as electrolysis. If the voltage applied
is such that the electrode reaction becomes diffusion limited the
current generated is directly proportional to the concentration.

Other electrochemical techniques are available; all, of course,
involve the reaction of the gas to be measured in some form of cell.

Again many of these methods are restricted to low concentration levels typical of atmospheric concentrations.

Summary

These notes have presented the basic operating principles of four main types of instrumental analysis, flame-ionization detectors, gas chromatographs, electromagnetic spectroscopy, and mass spectrometers. There are many other instrumental analysis techniques which have not been discussed here, as their use is still restricted to bench scale laboratory work. Some of these techniques, e.g. NMR and ESR spectroscopy, have already been used to obtain basic kinetic data on combustion reactions. They are discussed in some of the references. (1, 16)

A summary of the methods discussed and some of their basic features is given in the table on the next page.

PART II   APPLICATIONS TO ANALYSIS OF EMISSION SOURCES

The techniques discussed in the first part of these notes are not absolute measurement techniques; prior calibration of the instruments is necessary. Accurate calibration of the instruments is necessary. Accurate calibration and proper sampling procedures are two of the most important items to consider in designing a system for source analysis. These topics will be discussed first followed by discussion of various compounds and the methods of chemical analysis that can be used for them.

Calibration

Instrument calibration requires the preparation of accurate primary standards. Since most emission measurements are concerned with small concentrations the standards, which should be in the same concentration range, will be of low concentration and, hence, difficult to prepare. Many manufacturers provide calibration gases on order, the price of the gas increasing with the accuracy of the analysis provided. Large laboratories with accurate gas chromatographs or mass spectrometers can check the analysis provided by the supplier to assure that there have been no changes in gas concentration. It is useful to obtain several standards and check them against each other to avoid the possibility of obtaining a bad calibration gas. This is especially true if the standard gases are left in the cylinder for long periods of time where adsorption effects might become important.

It is also possible to prepare primary standards by flow techniques. Commercial devices are available which will provide accurate enough flow control to allow the mixing of two gases to make

SUMMARY OF GAS ANALYSIS TECHNIQUES IN AIR POLLUTION SOURCE ANALYSIS

Technique	Species	Sampling	Sensitivity	Response Times	Comments
Flame ionization	Hydrocarbons	Continuous or batch	ppb	Rapid	
Gas Chromatography	all	Batch		Slow	Applicable to wide range of gases
Thermal cond.			ppm		
Flame ionization			ppb		
Spectroscopy				Rapid	
Infrared	non-homopolar	Continuous or batch	10 ppm		Limited sensitivity
Visible/ultra-violet	all	Continuous or batch			Complex spectra
Mass Spectroscopy	all	Continuous or batch			
Field deflection			ppb	Rapid	Expensive
t. o. f.			10 ppm	Extremely rapid	Extremely rapid de-termination
Colorimetric	Special Applications	Continuous or batch	-	Generally slow	

a good standard. Thomas and Amtower (23) have reported on a portable
apparatus which uses gas dilution to obtain desired concentrations.
O'Keefe and Ortman (24) have discussed the use of permeation of
gases through plastic tubing to produce primary standards. Nelson
and Griggs (25) have described a device which uses a syringe whose
plunger is driven by a synchronous motor and a precision ratio gear-
box. They were able to obtain concentration ranges of 2 to 2000ppm
for evaporization of liquids in a carrier gas and 0.05 to 2000ppm
for dilution of a gas in the carrier, with an accuracy of $\pm$ 1%.

Sampling

The essential problem of sampling is to deliver a reliable sample to the analytical instrument; that is a sample whose species concentrations are the same when it reaches the instrument as they are
in the actual process being analyzed. Sampling techniques can be
broken down into the following classifications:

Grab samples: Small volumes of sample taken from the exhaust
stream. It is assumed that operating conditions are at steady state
during the sampling period.

Total (bag) sampling: The entire exhaust over a period of time
is collected to obtain a composite sample used to represent the
average behavior over the given set of operating conditions.

Proportional sampling: Similar to total sampling except that
only a fraction of the total exhaust stream is collected. The sample flow rate is varied so that the proportion of total exhaust
flow collected is constant during the entire sampling period.

Continuous sampling: A portion of the exhaust stream is sampled
for immediate analysis. The sample flow rate is kept constant. This
differs from the first 3 techniques which fall into the general
classification of batch sampling.

In all of these sampling techniques the potential sources for
losses are due to absorption on the walls, chemical reaction, and
trapping of compounds in condensate traps or reagents designed for
trapping other compounds.

The problem of adsorption on the walls of sampling devices can
be minimized by using a heated sampling system and using only stainless steel, teflon, or glass. Losses due to wall effects are particularly important in the analysis for oxides of nitrogen. It has
been found that some stainless steel containers and Tedlar bags have
minimal wall effects, while significant losses of nitrogen oxides
are found in others (26). Schuette (27) has published a note sum-
marizing experiences of different researchers with various plastic

bags for sample collection and storage.

As a general rule, then, sampling systems, particularly those used for analysis of oxides of nitrogen should be carefully checked with calibration gases to assure no losses due to wall effects prior to their uses for actual sampling.

The most significant loss due to homogenous reaction is due to reactions between oxides of nitrogen and hydrocarbons. These re-actions are, of course, similar to those occurring in the forma-tion of photochemical smog. They become especially important if samples must be stored for a long time between sampling and analysis or if the sample pressure (and therefore the reactant concentrations) are high. Stigsby et al.(28) noted a loss of olefinic hydrocarbons in stored samples. In particular, they found that higher molecular weight olefins disappeared almost completely after 100 minutes stan-ding time. The stability of hydrocarbon composition was improved if oxides of nitrogen were removed.

Dimitriades (26) has reported on a method for sampling exhaust gases to be analyzed for oxides of nitrogen in which the sample stream was passed over a Hopcalite catalyst at $1100^\circ C$ to remove hydrocarbons. Flow rates were adjusted so that the loss of NO was minimized. He found that results of NO measurements with the hydro-carbon removal were from 1 to 11 percent greater than similar me-thods without hydrocarbon removal.

Since the majority of the oxides of nitrogen in emissions, particularly automobile exhaust, is nitric oxide (NO) and the re-actions of hydrocarbons with oxides of nitrogen are primarily with nitrogen dioxide, it would seem that such reactions are not impor-tant. The problem is that with long storage times the NO can react with the excess oxygen in the exhaust gas to form NO_2. In addition it has been shown that the NO/O_2 reaction proceeds faster when hy-drocarbons are present (26). A further problem is that most methods of analysis for oxides of nitrogen involve an oxidation step where NO is converted to NO_2. The possibility of reactions with hydro-carbons becomes particularly important at these high NO_2 concentra-tions.

In order to avoid homogenous reactions, then, it is necessary to minimize sample holding time. When continuous sampling analysis is used, this presents no problem; if long storage times are required, it is advisable to take precautions to minimize reactions such as removal of hydrocarbons, if the primary interest is in nitrogen oxide analysis, or diluting the samples with an inert gas to mini-mize the concentrations and hence the reaction rates. Sampling systems can be checked for losses due to homogenous reactions by using mixtures of known concentrations of the compounds to be ana-

lyzed. In order to definitely attribute losses to homogenous reactions it is important to first verify that surface effects are
not important by checking the sampling system with the individual'
reactants.

Most sampling procedures remove water from the exhaust. This
is particularly necessary for analyses using gas chromatography
where the water can have harmful effects to the equipment. It is
also important in nitrogen oxides analysis where it has been found
that water vapor is always associated with losses of nitrogen oxides
(26). Typical condensation devices are ordinary heat exchangers
which can be simple glass laboratory condensers, if the sampling
rate is small. Also for small sampling rates various drying agents
can be used to remove trace quantities of water. In using condensate traps there is always the possibility that some of the emissions might be trapped out with the water. It has been found, however, that there is no appreciable loss of hydrocarbons due to condensate traps under normal conditions (29). There appears to be a
transient adsorption - through the condensers but can lead to erroneous results in a proportional sampling technique. Most drying
agents will adsorb hydrocarbons so that water removal prior to hydrocarbon analysis is usually restricted to condensation techniques.

It has been found that water traps retain less than 1% of the
total oxides of nitrogen in automobile exhaust streams (26). This
is probably due to the absorption of NO_2 and use of condensate traps
on streams with high concentrations of nitrogen dioxide would probably give low results. It has also been noted that the drying
agent $CaSO_4$ (commercial name: Drierite) does not adsorb nitric
oxide, but can adsorb NO_2. In nitric oxide analyses based on the
oxidation of NO to NO_2 this oxidation should be performed after
the sample has passed through the water collection system.

Since most exhaust gas streams contain particulate matter, it
is usually necessary to place a filter in the system to remove these
particles and avoid contamination of analytical instruments. This
is especially important in continuous flow instruments.

There is no sure sampling system that can be guaranteed to
deliver a representative sample to the analysis step. In designing
a sampling system it is important to carefully check it for possible
losses by using known mixtures under the same conditions that will
be used in the actual sampling process. This is especially true
in the development of new analytical methods.

Carbon Monoxide and Carbon Dioxide

The analysis of these gases is done almost exclusively by non-
dispersive infrared. The NDIR technique is especially suited to

these gases since they have strong absorption bands in regions where
there is little interference from other gases. If a laboratory
already has a gas chromatograph or dispersive infrared analyzer
these instruments can be readily used to measure CO and CO_2. The
NDIR, however, is probably the best instrument to use to set up
a new analytical facility for continual routine measurement of
these gases.

Jecko and Reynaud (30) have described an automatic gas chromatographic system for measurement of CO, CO_2 and H_2 in furnace stack
gases. The system uses two columns, a molecular sieve column to
separate out the CO and H_2 and a silica gel column for the CO_2. The
system is capable of taking a sample every 7 minutes.

Reckner, Scott, and Biller (31) used a dispersive spectrometer
to measure CO and CO_2, using the peaks at 4.6 and 13.5 microns, respectively. Fristrom and Westenberg (3) have described the data
reduction of a mass spectral analysis on Methane/Oxygen flame gases.
The present the simultaneous analysis of 8 species including CO and
CO_2.

Hydrocarbons

Hydrocarbon analyses can be made in terms of total hydrocarbons,
relative proportions in a given hydrocarbon class (i.e. olefins,
aromatics, paraffins), or on a compound by compound basis.

The two most common instruments for measurement of total hydrocarbons are the flame-ionization detector and the hexane sensitized
NDIR. Because of its earlier development the NDIR has become the
legal standard for vehicle emissions legislation. Jackson's study
(8) comparing the two methods has been discussed in the sections
on flame-ionization and nondispersive infrared. His study as well
as others show that the FID gives a better indication of the total
hydrocarbon content of an exhaust gas. This is particularly important since the hydrocarbons that the NDIR is not sensitive to,
the olefinic hydrocarbons, are the main contributors to photochemical
reactivity. Typically the total hydrocarbons measured by the flame-
ionization will be 1.8 times those measured by the nondispersive
infrared. Because of the common use of these two different devices
to measure total hydrocarbons one has to be careful to ascertain the
meaning of reported data on total hydrocarbon emissions; such data
are meaningless unless the analyzer is specified. Many researchers
prefer the FID since it gives a better picture of the total hydrocarbon emissions; people often measure total hydrocarbons by the
two methods and report both figures.

It is also possible to use a dispersive infrared analyzer to
obtain a measure of total hydrocarbons by measuring the absorption

at 3.4 microns (31). Many hydrocarbons have an absorption band in
this region due to C-H bond stretching vibrations. n-Hexane has a
particularly strong peak at 3.4 microns and is used as the calibration gas so that values are reported as equivalent n-Hexane.

Combustion of heavy fuels usually results in some high molecular weight hydrocarbons in the exhaust. It is necessary to maintain high temperatures in the analyzer to avoid adsorption of these
compounds. Pearsall (9) has described the construction of a high
temperature system, using a gas chromatograph flame-ionization detector, for measurement of total hydrocarbons in Diesel exhaust.

Klosterman and Sigsby (32) have developed a system which used
various trapping columns to make class identifications of exhaust
hydrocarbons with a flame ionization detector. They are able to
obtain a value for the concentration of total hydrocarbons, olefins
(plus acetylene), alkylbenzenes, and paraffins (plus benzene).

Because of the large number of hydrocarbons present in exhaust
gases the gas chromatograph is the only way that these compounds
can be separated, identified, and measured on an individual basis.
Several different columns and operating conditions have been proposed for analysis of exhaust gas hydrocarbons; all of them use a
flame-ionization detector.

Papa, Dinsel, and Harris (33,34) have described a chromatographic system which can determine 200 hydrocarbons in the C_1-C_{12}
range. Their method uses a peaked column for the C_1 and C_2 hydrocarbons and a capillary column for C_3-C_{12} hydrocarbons. The system
is capable of detecting concentrations of one part per billion for
each compound. A sample chromatogram and a table identifying the
peaks taken from reference 34, are given on the next two pages.

McEwen (35) has described a system for chromatographic analysis of automobile exhaust which uses a trapping column and two
analytical columns. A mercuric perchlorate trapping column was
used to remove all compounds except paraffins; two separate analyses
are required, one for paraffins only and one for all components.
The unsaturated hydrocarbons are found by difference.

Jacobs (36) used a single capillary column to separate and
identify 85 different hydrocarbons in the C_1 to C_{10} range. An
analysis which can identify concentrations of 1 part per million can
be completed in 13 minutes.

Moore, Katz, and Drowley (37) used columns packed with alumina
coated with cyclohexane and ether to separate polynuclear aromatic
compounds. They were able to obtain effective separation between
several of these compounds.

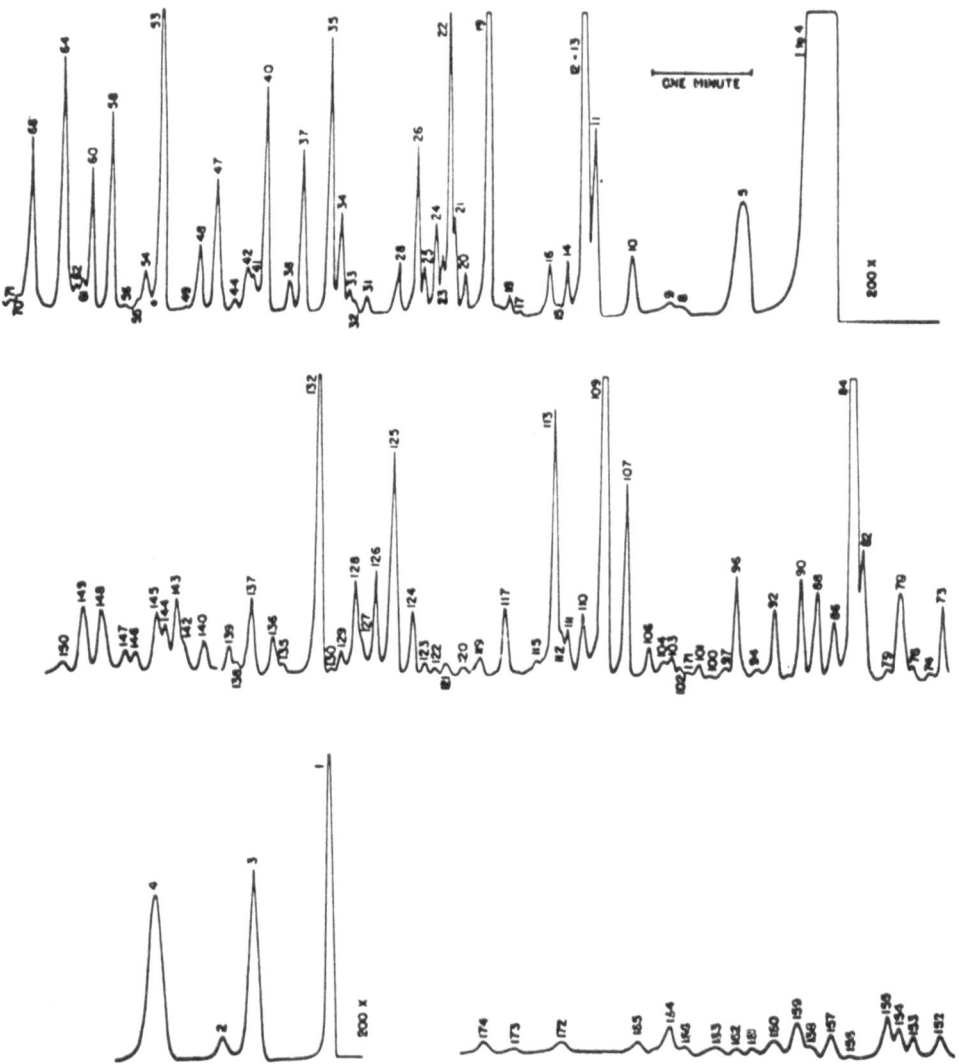

Figure 15.   Chromatogram of exhaust hydrocarbons.   (Identification below)

IDENTIFICATION OF HYDROCARBON PEAKS

Peak Number	Hydrocarbons	Peak Number	Hydrocarbons	Peak Number	Hydrocarbons
1	Methane	4	Acetylene	7	Cyclopropane
2	Ethane	5	Propylene	8	Propadiene
3	Ethylene	6	Propane	9	Methylacetylene

IDENTIFICATION OF HYDROCARBON PEAKS (continued)

Peak Number	Hydrocarbons	Peak Number	Hydrocarbons	Peak Number	Hydrocarbons
10	Isobutane	39	2-Ethyl-1-butene	61	Unknown 61
11	Isobutylene and/or 1-Butene	40	n-Hexane	62	Unknown 62
		41	trans-3-Hexene	63	Unknown 63
12	1,3-Butadiene	42	trans-2-Hexene	64	2,2,4-Trimethyl-pentane
13	n-Butane	43	2-Methyl-2-pentene	65	1-Heptene
14	trans-2-Butene	44	cis-3-Hexene	66	Unknown 66
15	Unknown 15			67	trans-3-Heptene
16	cis-2-Butene	45	cis-2-Hexene	68	n-Heptane
17	Unknown 17			69	cis-3-Heptene and/or 3-Ethyl-trans-2-pentene
18	3-Methyl-1-butene	46	3-Methyl-trans-2-pentene and/or 3-Methyl-cis-2-pentene	70	2,4,4-Trimethyl-1-pentene and/or trans-2-Heptene
19	Isopentane			71	cis-2-Heptene
20	1-Pentene			72	2,5-Dimethyl-trans-3-hexane
21	2-Methyl-1-butene	47	Methylcyclo-pentane	73	Methylcyclohex-ane
22	n-Pentane	48	2,4-Dimethyl-pentane	74	Unknown 74
23	2-Methyl-1,3-butadiene	49	2,2,3-Trimethyl-butane	75	Unknown 75
				76	2,4,4-Trimethyl-2-pentene
24	trans-2-Pentene	50	3,4-Dimethyl-1-pentene	77	4-Methyl-1-cyclo-hexene
25	cis-2-Pentene	51	4,4-Dimethyl-cis-2-pentene		
26	2-Methyl-2-butene	52	3,3-Dimethyl-pentane	78	2,4-Dimethylhex-ane and/or 2,5-Dimethylhexane
27	Unknown 27	53	Benzene	79	2,2,3-Trimethyl-pentane
28	2,2-Dimethylbu-tane	54	Cyclohexane	80	Unknown 80
29	Unknown 29	55	3-Ethyl-1-pentene	81	4-Methylheptane
30	Unknown 30	56	5-Methyl-1-hexene	82	2,3,4-Trimethyl-pentane
31	Cyclopentene	57	4-Methyl-1-hexene	83	Unknown 83
32	4-Methyl-1-pentene and/or 3-Methyl-1-pentene	58	2-Methylhex-ane and/or 2,3-Dimethylpen-pentane	84	Toluene
				85	Unknown 85
33	Cyclopentane			86	2,3,3-Trimethyl-pentane
34	2,3-Dimethyl-butane	59	Cyclohexene	87	2,5-Dimethyl-trans-2-hexene
35	2-Methylpen-tane	60	3-Methylhexane	88	2-Methyl-3-ethylpentane and/or 2,3-Dimethyl-hexane
36	4-Methyl-cis-2-pentene				
37	3-Methylpentane				
38	2-Methyl-1-pentene and/or 1-Hexene				

IDENTIFICATION OF HYDROCARBON PEAKS (continued)

Peak Number	Hydrocarbons	Peak Number	Hydrocarbons	Peak Number	Hydrocarbons
89	Unknown 89	124	n-Propylbenzene	147-	
90	3,4-Dimethylhexane and/or 3-Methylheptane	125	1-Methyl-4-ethylbenzene- and/or 1-Methyl-3-ethylbenzene	153	Unknown
				154	Durene
				155-	
91	Unknown 91	126	1,3,5-Trimethylbenzene	172	Unknown
92	2,2,5-Trimethylhexane			173	1-Dodecene
		127	Unknown 127	174-	
93	1-Octene	128	1-Methyl-2-ethylbenzene	200	Unknown
94	trans-1,2-Dimethylcyclohexane	129	Unknown 129		
95	Unknown 95	130	Unknown 130		
96	n-Octane	131	t-Butylbenzene		
97	trans-2-Octene	132	1,2,4-Trimethylbenzene		
98	Unknown 98				
99	Dimethylheptane	133	Unknown 133		
		134	Isobutylbenzene		
100	cis-2-Octene	135	Unknown 135		
101	cis-1,2-Dimethylcyclohexane	136	Unknown 136		
		137	sec-Butylbenzene		
102	Unknown 102				
103	Ethylcyclohexane	138	1-Methyl-3-isopropylbenzene		
104	Unknown 104				
105	Unknown 105				
106	Unknown 106	139	n-Decane		
107	Ethylbenzene	140	1,2,3-Trimethylbenzene		
108	Unknown 108				
109	m + p-Xylene	141	1-Methyl-4-ispropylbenzene		
110	Unknown 110				
111	Unknown 111				
112	Unknown 112	142	1,3-Diethylbenzene		
113	o-Xylene				
114	2-Methyloctane	143	Unknown 143		
115	Unknown 115	144	n-Butylbenzene and/or 1-Methyl-4-n-propylbenzene		
116	Unknown 116				
117	n-Nonane				
118	Unknown 118				
119	Isopropylbenzene	145	1,3-Dimethyl-5-ethylbenzene and/or 1,2-Diethylbenzene		
120	Unknown 120				
121	Unknown 121				
122	Unknown 122	146	1-Methyl-2-n-propylbenzene		
123	Unknown 123				

Oxides of Nitrogen

A review of the various methods for determining the concentrations of nitrogen oxides in auto exhausts has been given by Dimitriades (38). The two most common oxides of nitrogen found in exhaust streams are nitric oxides, NO, and nitrogen dioxide, NO_2. Often the total concentration of these two compounds is reported as NO_x. The NO is formed during the high temperature combustion process; the subsequent oxidation to NO_2 is a relatively slow reaction which usually takes place only in exhaust streams with large concentrations of NO and large amounts of excess oxygen. Campuau and Neerman (19) used a mass spectrometer to separately measure the NO and NO_2 concentration in automobile exhausts. They found that 99% of the total oxides of nitrogen was NO. Studies on Diesel exhaust, (31) however, have shown that 37% of the total nitrogen oxides are NO_2 under idling conditions (which give large excess air in a Diesel).

For streams which are known to contain only NO the nondispersive infrared NO analyzer is the best analyzer from the standpoint of ease, accuracy, and response time (38). For measurements of NO_2 a nondispersive ultraviolet analyzer can be easily used as the familiar brown color of NO_2 absorbs strongly in the visible and ultraviolet. The construction of the nondispersive ultraviolet (NDUV) analyzer is relatively simply and commercial instruments are available.

The NDUV can also be used to measure total oxides of nitrogen by taking a batch sample and adding oxygen to oxidize the NO to NO_2. This method was developed by Nicksie and Harkins (39) and is discussed by Dimitriades (26,38). By taking two measurements, one of the direct sample and one of the oxidized sample, one obtains a measure of the NO_2 and the total NO + NO_2. The NO can then be found by the difference. The oxidation step requires about 15 minutes. This long delay and the extreme reactivity of NO_2, which requires precautions in carrying out the analysis are the main difficulties with this method. Singh et al. (40) have described a modification of this technique which uses ozone as the oxidant. They were able to develop a continuous system with response times less than one minute, but found interference from hydrocarbons in the exhaust.

Newhall and Starkman (41) used the 5.3 microns emission band of NO, suitably corrected for interferences, to follow the nitric oxide concentration during the expansion stroke of a single cylinder engine. Other work (31) used the 5.25 and 6.16 micron bands to measure NO and NO_2, respectively, in Diesel exhaust.

Systems for the chromatographic separation and analysis of nitrogen oxides have been described, (42,43) but these are usually difficult to operate as columns must be extensively pretreated to avoid severe tailing.

The oxides of nitrogen in source effluents can also be measured by modifications of the colorimetric techniques used for atmospheric analysis. There are two basic reagents, phenyldisolfonic
acid and Saltzmann solution (44).

Oxygenated Hydrocarbons

The determination of these compounds involves collection of
the compounds in an appropriate reagent for subsequent gravimetric,
colorimetric, or chromatographic analysis. A common reagent used
to collect the carbonyl compounds is 2,4-dinitrophenylhydrazine
(2,4-DNPH). The aldehydes and ketones are converted to 2,4-dini-
trophenylhydrazone derivatives. Oberdorfer, (45) describes a
procedure where these derivatives are precipitated, collected, and
weighed to obtain a measure of total carbonyls as equivalent formaldehyde. Papa (46) has modified the collection reagent by using a
70% pyradine medium which stabilizes the color of the product. He
is thus able to obtain a colorimetric measure of the total carbon-
lyls.

Fracchia, Schuette, and Mueller (47) describe a method where
the 2,4-DNPH derivatives are collected by filtration and extraction,
dissolved in carbon disulfide, and analyzed by gas chromatography.
Barber and Sawicki (48) have developed a method using 4-nitrophenyl-
hydrzone and paper or thin layer chromatography to analyze aromatic
arbonyl compounds in exhausts.

Levaggi and Feldstein (49) have developed a method for analysis
of aldehydes from industrial surface coating, drying, and baking
ovens. They collected the sample gas in a bisulfite solution and
used a chromatographic analysis to measure the concentrations of
low molecular weight carbonyls. Their method was not able to measure
formaldehyde chromatographically but this could be determined colorimetrically.

Sulfur Oxides

Because of the corrosive nature of these gases, it is especially
important to have a well tested sampling system. It is necessary
to remove water to avoid the formation of sulfuric acid which not
only gives a bad analysis but severely damages analyzers.

It is possible to use nondispersive infrared analysis on SO_2
if the sample is dried. Reckner, Scott, and Biller (31) were not
able to obtain accurate analysis of SO_2 by dispersive infrared analysis at 7.25 microns because of water interference. They used a
colorimetric method to analyze SO_2. Low and Clancy (50) have proposed the use of dispersive spectroscopy for the remote measurement
of stack gas concentrations. Their preliminary work indicated quali-

tative distinctions of SO_2 and CO_2.

Two gas chromatographic systems for separation and identification of SO_2 have been described. (51,52) Such analyses are difficult because the SO_2 tends to irreversibly adsorb on some column materials and preconditioning is necessary.

Smith, Hultz, and Orning (53) have prepared a report on sampling and analysis of sulfur oxides and nitrogen oxides from large coal-fired steam generators.

Summary

The references given here constitute only a small fraction of the many articles in the literature on the analysis of various pollutants. Well established methods are available for the analysis of common compounds, and many papers have been published for various special analyses.

In any case the problems of sampling and standardization must be carefully considered.

BIBLIOGRAPHY

1. Pecsok, R. L., and L. D. Shields, Modern Methods of Chemical Analysis, Wiley, 1968.

2. Bair, E. J., Introduction to Chemical Instrumentation, McGraw-Hill, 1962.

3. Fristrom, R. M., and A. A. Westenberg, Flame Structure, New York, McGraw-Hill, 1965, Chapters 7-11.

4. Altshuller, A.P., "Air Pollution", Anal. Chem. 41:1R-13R (Analytical Review Issue - April, 1969).

5. Altshuller, A. P., "Air Pollution," Anal. Chem. 39: 10R-21R (Analytical Review Issue - April, 1967).

6. McWilliam, I. G. and R. A. Dewar, in "Gas Chromatography" (D.H. Desty, ed.) Volume 2, page 142, Butterworths, London, 1958.

7. Ackman, R. G., "The Flame Ionization Detector: Further Comments on Molecular Breakdown and Fundamental Group Responses," J. Gas Chromtog. 6:497-501 (Oct., 1968).

8. Jackson, M. W., "Analysis for Exhaust Gas Hydrocarbons --

Nondispersive Infrared Versus Flame-Ionization," J. Air Poll. Control Assoc. 11: 697-702 (December, 1966)

9. Pearsall, H. W., "Measuring the Total Hydrocarbons in Diesel Exhaust", Preprint No. 670089 SAE Automotive Engineering Congress, Detroit, Jan 9-13, 1967.

10. Kaiser, R., Gas Phase Chromatography, P. H. Scott, Translator, Washington, Butterworths, 1963.

11. Knox, J. H., Gas Chromatography, London, Methuen, 1962.

12. Hoffman, R. L. and C. D. Evans, "Thermal Conductivity Effect of Carrier Gases on Flame-Ionization Detector Sensitivity," Science, 153: 172-173 (8 July 1966).

13. Hoffman, R. L. and C. D. Evans, "Gas-Solid Chromatography of Hydrocarbons on Activated Alumina," Anal. Chem. 38: 1309-1312 September, 1966).

14. Lovelock, J. E., "Ionization Methods for the Analysis for Gases and Vapors," Anal. Chem., 33:162-176 (February, 1961).

15. Schwartz, R. D., D. J. Brasseaux, and R. G. Mathews, "High-Resolution Capillary Adsorption Columns for Gas Chromatography," Anal. Chem. 38: 303-306 (February, 1966).

16. Whiffen, D. H., Spectroscopy, Wiley, 1966.

17. Sawyer, R. F., "The Homogenous Gas Phase Kinetics of Reactions in The Hydrazine - Nitrogen Tetroxide Propellant System," Ph. D. Thesis in Engineering, Princeton, 1965.

18. Kiser, R. W., Introduction to Mass Spectrometry and its Applications, Prentice-Hall, 1965.

19. Campau, R. M. and J. C. Neerman, "Continuous Mass Spectrometric Determination of Nitric Oxide in Automotive Exhaust," SAE Trans. 75:582-591 (1967).

20. McFadden, W. H., "Introduction of Gas Chromatographic Samples to a Mass Spectrometer," Separation Science 1:723-746 (1966).

21. Burson, K. R. and C. T. Turner, "A Simple Trap for Collecting Gas Chromatographic Fractions for Mass Spectometer Analysis" J. Chromatog. Sci. 7:63-64 (Jan. 1969).

22. Black, D. R., R. A. Flath, and R. Teranishi, "Membrane Molecular Separators for Gas Chromatographic - Mass Spectrometric

Interfaces," J. Chromatog. Sci. 7: 284-289 (May, 1969).

23. Thomas, M. D. and R. E. Amtower, "Gas Dilution Apparatus for Preparing Reproduceable Dynamic Gas Mixtures in Any Desired Concentration and Complexity," J. Air Pollution Control Assoc., 16: 618-623 (November, 1966).

24. O'Keefe, A. E. and G. C. Ortman, "Primary Standards for Trace Gas Analysis," Anal. Chem., 38 760-763 (May, 1966).

25. Nelson, G. O., and K. S. Griggs, "Precision Dynamic Method for Producing Known Concentrations of Gas and Solvent Vapor in Air," Rev. Sci. Inst. 39:927-928 (June, 1968).

26. Dimitriades, B., "Determination of Nitrogen Oxides in Auto Exhaust," J. Air Poll. Control Assoc., 17: 238-243 (April, 1967).

27. Schuette, F. J., "Plastic Bags for Collection of Gas Samplers," Atmospheric Environment 1:515-519 (July, 1967).

28. Sigsby, J. E., T. A. Bellar, M. L. Bellar, and D. L. Klosterman, 150th National Meeting, American Chemical Society, Atlantic City, N. J., September 12-17, 1965. (Cited on page 15R of reference 5).

29. Fleming, R. D., B. Dimitriades, and R. W. Hurn, "Procedures in Sampling and Handling Auto Exhaust," J. Air. Poll. Control Assoc., 15: 371-374 (August, 1965).

30. Jecko, G. and B. Reynaud, "Analysis des gaz de haut forneau par chromatographie gaz-solide" (Analysis of Stack Gases by Gas-Solid Chromatography), Rev. Met. (Paris) 64: 681-686 (July-August, 1967).

31. Reckner, L. R., W. E. Scott and W. F. Biller, "The Composition and Odor of Diesel Exhaust", API Proceedings, Division of Refining 45 (III):133-147 (1965).

32. Klosterman, D. L. and J. E. Sigsby, Jr., "Application of Subtractive Techniques to the Analysis of Automotive Exhaust," Environmental Sci. & Tech., 1: 309-314 (April, 1967).

33. Papa, L. J., D. L. Dinsel, and W. C. Harris, "Gas Chromatographic Determination of C_1 to C_{12} Hydrocarbons in Automotive Exhaust," J. Gas Chromatog. 6:270-279 (May, 1968).

34. Papa, L. J., "Gas Chromatography - Measuring Exhaust Hydrocarbons Down to Parts per Billion", SAE Trans., 76:1797-1819, (1968).

35. McEwen, D. J., "Automobile Exhaust Hydrocarbon Analysis by Gas Chromatography," Anal. Chem. 38: 1047-1053 (July, 1966).

36. Jacobs, E. S., "Rapid Gas Chromatographic Determination of C_1-C_{10} Hydrocarbons in Automotive Exhaust Gas," Anal. Chem. 38: 43-48 (January, 1966).

37. Moore, G. E., M. Katz, and W. B. Drowley, "Polynuclear Aromatic Hydrocarbons in Urban Atmospheres in Ontario," J. Air Poll. Control Assoc. 16: 492-497 (September, 1966).

38. Dimitriades, B., "Methods for Determining Nitrogen Oxides in Automotive Exhausts," U.S. Bureau of Mines RI 7133, May, 1968.

39. Nicksic, S. W., and J. Harkins, "Spectrophotometric Determination of Nitric Oxide in Auto Exhaust," Anal. Chem., 34: 985-988 (1962).

40. Singh, T., R. F. Sawyer, E. S. Starkman, and L. S. Caretto, "Rapid Continuous Determination of Nitric Oxide in Exhaust Gases," J. Air Poll. Control Assoc. 18: 102-105 (Feb., 1968).

41. Newhall, H. K., and E. S. Starkman, "Direct Spectroscopic Determination of Nitric Oxide in Reciprocating Engine Cylinders", SAE. Trans 76: 743-776 (1968).

42. Dietz, R. N., "Gas Chromatographic Determination of Nitric Oxide on Treated Molecular Sieve", Anal. Chem. 40 1576-1578 (August, 1968).

43. Trowell, J. M., "Gas Chromatographic Separation of Oxides of Nitrogen," Anal. Chem. 37: 1152-1154 (August, 1965).

44. Davis, R. F., and W. E. O'Neill, "Determination of the Oxides of Nitrogen in Diesel Exhaust Gas by a Modified Saltzman Method," Bureau of Mines RI 6790, 1966.

45. Oberdorfer, P.E., "The Determination of Aldehydes in Automotive Exhaust Gases," SAE Trans. 76: 743-762 (1968).

46. Papa, L. J., "Colorimetric Determination of Carbonyl Compounds in Automotive Exhaust as 2,4-Dinitrophenylhydrazones", Environ. Sci. Technol. 3:397-398 (April, 1969).

47. Fracchia, M. F., F. J. Schuette, and P. K. Mueller, "A Method for Sampling and Determination of Organic Carbonyl Compounds in Automobile Exhaust," Environ. Sci. Technol. 1: 915-922 (Nov., 1967).

48. Barber, E. D. and E. Sawicki, "Separation and Identification
     of Aromatic Carbonyl Compounds as their 4-Nitrophenylhydrazones
     by Paper and Thin Layer Chromatography", Anal. Chem. 40: 984-
     986 (May, 1968).

49. Levaggi, D.A. and M. Feldstein, "The Collection and Analysis of
     Low Molecular Weight Carbonyl Compounds from Source Effluents",
     J. Air Poll. Control Assn. 19: 43-45 (Jan., 1969).

50. Low, M.J.D., and Clancy, F. K., "Remote Sensing and Characteri-
     zation of Stack Gases by Infrared Spectroscopy. An approach
     using Multiple-Scan Interferometry," Environmental Science
     Technology 1: 73-74 (Jan., 1967).

51. Obermiller, E. L., and G. O. Charlier, "Gas Chromatographic
     Separation of Nitrogen, Oxygen, Argon, Carbon Monoxide, Hy-
     drogen Sulfide, and Sulfur Dioxide," J. Gas Chromatog. 6: 446-
     447 (Aug., 1968).

52. Kopke, R. K., and D. F. Adams, "Evaluation of Gas Chromatogra-
     phic Columns for Analysis of Subparts per Million Concentra-
     tions of Gaseous Sulfur Compounds," Environ. Sci. Technol. 1:
     479-481 (June, 1967).

53. Smith, J. F., J. A. Hultz, and A. A. Orning, "Sampling and
     Analysis of Flue Gas for Oxides of Sulfur and Nitrogen," U.S.
     Bureau of Mines, RI 7108 (April, 1968).

VEHICULAR EMISSIONS AND CONTROL

Ernest S. Starkman

University of California

Berkeley, California

Introduction

Internal Combustion Engines, now, and for the immediate future, constitute the principal and preponderant contributors to atmospheric pollution. Battery powered or direct energy conversion vehicles may become an entity at some future date. However, for at least a decade, limitations in their application will preclude any significant replacement of combustion engines for vehicular propulsion.

The piston engine now dominates in surface transport application. Spark ignition engines, burning gasoline, are almost exclusively the powerplant for passenger cars and light trucks. Diesel engines power about 60 per cent of the trucks weighing over 33,000 pounds gross, and a larger proportion of buses. There are now almost 100 million combustion engine powered vehicles registered in the United States. [1] Fifteen million are trucks and buses. The world total is approximately 200 million passenger cars, trucks and buses. The gasoline engine has come under the scrutiny of pollution control agencies first, but the Diesel engine is also commencing to receive attention.

Gas turbines already constitute the predominant powerplant for aircraft propulsion and will soon enter into heavy truck applications. There is indication that the gas turbine may also be a serious contender in the passenger car area within a decade. Other possibilities, such as Wankel engines or free piston engines, external combustion and steam engines do not appear to be in contention. The material presented below will therefore be limited to the principles which determine the operation and emission production from the three pre-

dominant engine types, spark-ignition, Diesel and gas turbine, and
most attention will be given to the spark ignition engine.

In presenting this material it should be pointed out that non-
vehicular engines also constitute a significant entity in pollution
considerations. About 7 million outboard engines are in use at the
present time, and about 8 million engines which were not for automotive, aircraft or boat applications were produced and sold in the
U. S. alone in 1966. (The total number of these non-propulsion
engines in present use is a figure not readily available.) Of the
latter, 94 per cent are one cylinder engines, however.[1] These
were for such applications as lawn movers, chain saws and similar
applications. Between the out-boards and the other non-automotive
installations, the use factor and total fuel consumption is presently
a relatively lesser entity in the total contribution to air pollution. But, as controls are effected on vehicular engines, the contributions from these other sources will take on increasing importance. In any event, the principles which apply to vehicular engine
combustion processes are equally applicable to non-vehicular.

The material which follows is a presentation of principles of
operation, source and character of pollutants from engines. It is
also concerned with devices and systems for control of vehicular
emissions and possibilities for alternative means for propulsion.

I. ENGINE TYPES AND THEIR CHARACTERISTICS

Internal Combustion Engine Principles

It will be assumed that sufficient general familiarity exists
with respect to the structural and mechanical features of engines
that there is no need to go into detail here. It should be necessary
only to point out that the piston engine operates by virtue of the
entrapment of successive cylinders full of fuel-air mixture. These
are compresssed, ignited, expanded and exhausted. In the gas turbine the compression is accomplished by rotary means. Fuel is added
to the compressed air and burned. The hot gases are then expanded
in a turbine and exhausted to the atmosphere.

More detail on engine principles and characteristics are available from textbooks on the subject. (2,3)

A. Piston Engines

1. The Spark Ignition Engine - Otto Cycle

All spark ignition engines operate on the same cycle, suggested
in France in 1862 by Beau de Rochas but credited to Otto, who built
a successful operating machine in 1876. The cycle is shown ideally

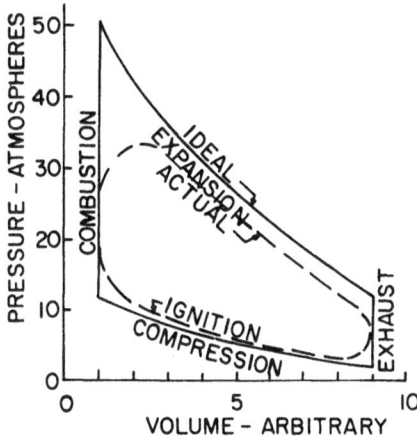

Figure 1. Otto Cycle. Compression Ratio of 9:1

in Figure 1. An indication of the order of magnitude of pressures
and temperatures which can be encountered is incorporated in this
figure.

Note that the actual pressure obtained at peak conditions is
about 30 to 40 atmospheres absolute and the temperature about 4500°R
to 5000°R. The pressure at exhaust is usually well above atmospheric
and the temperature about 1500°R to 2000°R.

Fuels for spark engines are almost entirely gasoline, with hydrocarbons ranging from C_4 to C_{14}. Even the few available directly injected spark engines usually use the same gasoline as carbureted
versions. Fuel-to-air ratios are almost always between the limits of
10 per cent deficiency of fuel, or lean, to the 40 per cent or rich,
of chemically correct. As will be shown, production of potential
air pollutants is very much dependent upon fuel-air ratios, spark
timing and compression ratio, among other items.

2. The Compression Ignition Engine - Diesel Cycle

The cycle proposed and developed by Rudolf Diesel incorporates
compression of air only and through the temperatures created thereby, self-ignition of subsequently injected fuel. The cycle is shown
in Figure 2.

Engine output is controlled by the amount of fuel injected
during each cycle. Fuel-to-air ratios range from less than 10 per

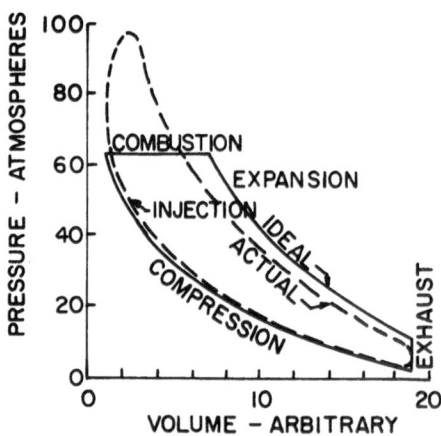

Figure 2. Diesel Cycle. Compression Ratio of 19:1

cent to approximately 70 percent of stoichiometric. Thus there is
always excess air. Actual temperatures depend upon the quantity
of fuel injected as do pressures, to a large extent. Because of the
higher compression ratios used, the pressures developed in Diesel
engines can be higher than a spark ignition, but the temperatures
usually do not exceed those in the spark engine.

The resulting principal differences between Diesel and spark
engines can be enumerated as:

1. Much higher compression ratios are used in the Diesel engine (14 to 22, average 18) than in gasoline (6 to 11, average).
2. The spark engine uses a premixed combination of fuel and
air which is compressed before ignition while the Diesel uses injection of fuel, after the air has already been compressed.
3. The Diesel engine uses self-ignition while the spark engine uses timed electrical discharge for ignition.

B. The Gas Turbine - Brayton Cycle

The evolution of the principles and the cycle on which the gas
turbine operates predate those of the piston engine, but the development to practice has been slower. Brayton generally receives recognition for the thermodynamic cycle, shown in Figure 3.

Fuel-to-air ratios presently are limited by the maximum temperatures which available materials for construction of the turbine
blades can withstand. These temperatures are now in the 2250°R to
2600°R range. Corresponding fuel-to-air ratios are from about 0.05

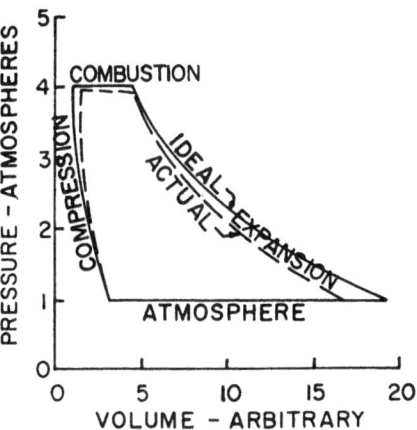

Figure 3. Brayton Cycle. Pressure Ratio 4:1

of chemically correct at idle to about 0.25 at maximum load. Thus
there is a considerable excess of air under all operating conditions.

II. NATURE OF EMISSIONS FROM ENGINES

Potential air pollutants from motor vehicles can be introduced into the atmosphere from the exhaust, the crankcase vent, the
carburetor and the fuel tank. Those materials which are produced
by the reciprocating engine as a consequence of the combustion process can contain all of the following:

A. Unburned hydrocarbons
B. Carbon monoxide
C. Oxides of nitrogen
D. Carcinogens
E. Particulate matter (smoke)
F. Lead
G. Odor
H. Oxides of sulfur

A. Unburned Hydrocarbons

Increasingly frequent instances of eye irritation, atmospheric
occlusion and vegetation damage in the Los Angeles basin during the
late 1940's were the principal motivating factors for research which
resulted in the proposal that a photosynthesized chemical reaction
was the responsible mechanism. This concept, generally credited
to Haagen-Smit[4], established that unburned hydrocarbons and oxides

of nitrogen in the presence of ultraviolet radiation, were mainly responsible for the irritating and damaging consequences of the so-called "smog" in Los Angeles.

Further work has disclosed that hydrocarbons and their derivatives vary in reactivity in this mechanism. In a representation of papers on this subject, (5 through 13) it is concluded that the tendency to react is very much a function of chemical structure.

Generally the spark engine in its uncontrolled form and in a representative driving pattern (14) produces from 100 to 1500 parts per million of unburned hydrocarbon. The idle and deceleration periods are the most productive of these materials.

It is generally agreed that flame quenching at the combustion chamber wall and in cracks and crevices, is the principal mechanism which produces unburned hydrocarbons in the exhaust of a spark ignition engine. (15) Thus, any factors which either contribute to

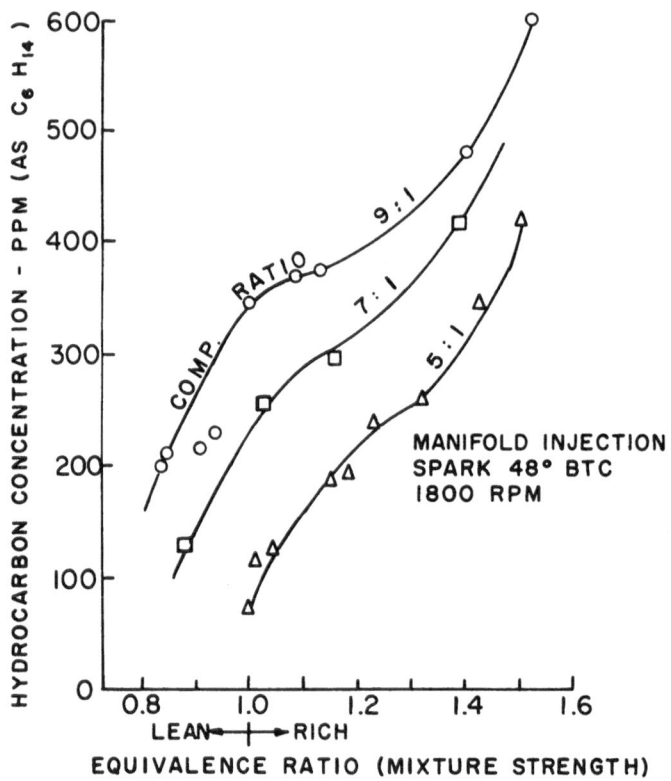

Figure 4.   Influence of Compression Ratio and Mixture Ratio on Exhaust Hydrocarbons. Single Cylinder Engine

the amount of fuel at the wall, or which influence the thickness of
the quench zone will as a consequence influence the quantity of
hydrocarbons emitted in the exhaust.

The concentration and chemical character of unburned hydrocarbons resulting from engine combustion are extremely variable. Almost
every possible design factor and operating condition is involved.
Additional effects can be obtained by varying the fuel characteristics.

Examples of the comparative levels of unburned hydrocarbon from
a single cylinder engine operating at a series of compression ratios
is shown in Figure 4. These data were taken at the University of
California, Berkeley. All other factors were identical except for
the variables shown.

The trend in concentration of unburned hydrocarbons in Figure
4 is largely reflective of the variation in thickness of the quench
zone and the fuel which is contained in that layer of combustion
chamber gases. It can be concluded that the minimum amount of unburned hydrocarbon to be obtained from a given homogeneous mixture
engine is that which is the quench zone.

A great deal of study has been expended on these various factors,
ranging from carburetion control to the effect of deposits and of
combustion chamber surface area. Representative of this work are
references 16 through 20.

One of the routes to overall hydrocarbons emission reduction,
considering only the fuel-air mixture, involves control of mixture
ratio to a value as lean as is consistent with acceptable performance, (17) restraint of fuel drops which might be thrown to the
walls by centrifugal action and possible fuel shut-off during deceleration.

The Diesel engine exhaust contains much less unburned material
than the spark ignition, but what is emitted is generally more un-
pleasant from the standpoint of odor. The principal objectionable
materials appear to be oxygenates.

The gas turbine shows particular promise as a low emitter of
hydrocarbons. Typical values from vehicular turbines now being
developed are quoted as on the order of a few parts per million.
When corrected for the effect of excess air, these numbers become
50 to 100 parts per million. Some typical values from aircraft and
small stationary turbines are shown in Figures 5 and 6, taken from
references 21 and 42.

B.   Carbon Monoxide

Carbon monoxide concentration tends to follow directly the fuel-

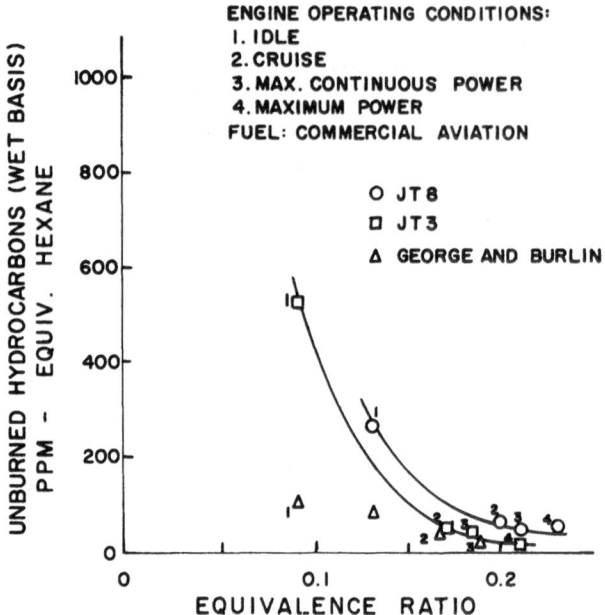

Figure 5.   Commercial Jet Engine Exhaust Hydrocarbon Content.
(Corrected for excess air to $\phi + 1.0$)

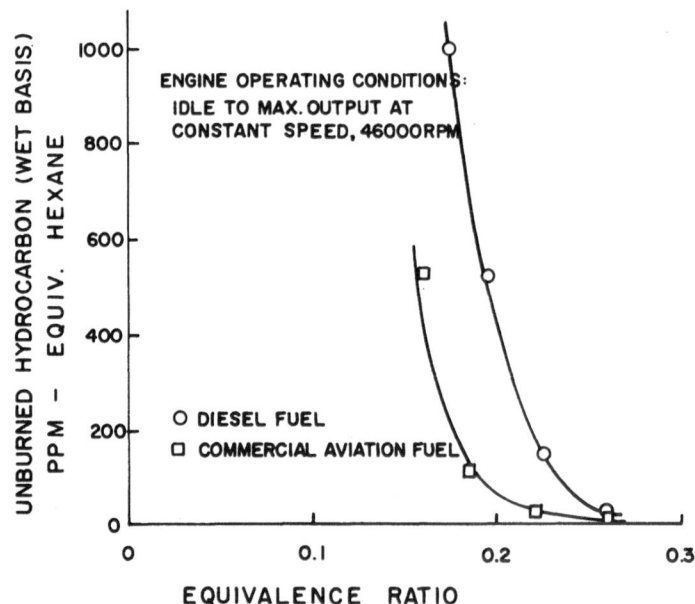

Figure 6.   Small Gas Turbine Exhaust Hydrocarbon Content.
(Corrected for excess air to $\phi = 1.0$)

air ratio. Leaning of the mixture results in proportionate reductions in carbon monoxide in all of the studies which report on this phenomenon.

Attention is being given to the non-equilibrium characteristics of the amounts of carbon monoxide and of nitric oxide which appear in the exhaust. (22) Typical of such non-equilibrium, as example, are the data for concentrations of carbon dioxide, carbon monoxide, water and hydrogen which are measured in the engine exhaust.

The most simple test to apply in order to determine how far the concentration of exhaust gases is from equilibrium is to insert into the water-gas equilibrium constant relationship,

$$K_p = \frac{(H_2O)\,(CO)}{(H_2)\,(CO_2)}$$

the concentrations of gases which are actually measured. Such a simple test yields equilibrium constants (K_p) which are between 4.1 and 4.5. This corresponds to an equilibrium temperature of about $3600°R$ ($2000°K$) rather than the approximately $1800°R$ ($1000°K$) actual temperature of the gases as they exit from the engine.

Another example of the non-equilibrium nature of carbon monoxide concentration is shown in Figure 7.

It can be seen that the amount of carbon monoxide produced by an engine at any mixture strength is always greater than that calculated for the moment that the exhaust valve opens. Particularly at very lean mixtures, the carbon monoxide concentration is many times more than it should be if it were in equilibrium.

The present explanation for the presence of greater than equilibrium quantities of carbon monoxide is that the reaction rate for conversion of CO to CO_2 is a relatively slow one except at very high temperatures. Thus the carbon monoxide is "frozen" at a value corresponding to very high temperature and very little piston travel after peak cylinder conditions.

Carbon monoxide levels from Diesel engines are so low as to be no problem. Similarly, the concentrations from gas turbines are extremely low. (21)

C. Oxides of Nitrogen

There is presently much dispute over the importance those oxides of nitrogen which are produced by an engine contribute to photochemical smog. There is agreement that oxides of nitrogen are necessary

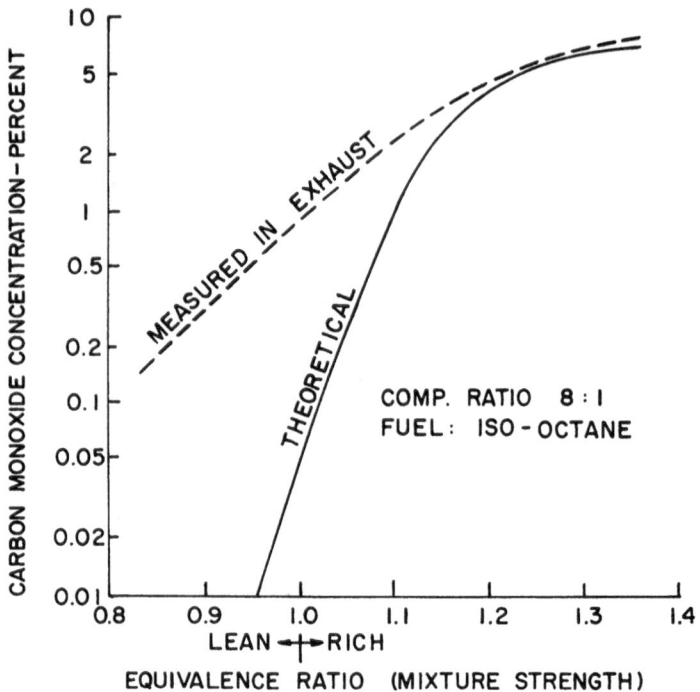

Figure 7.   Theoretical and Actual Quantities of Carbon Monoxide in
Spark Ignition Engine Exhaust.

for the photochemical process which produces the more harmful materials which result.

There is even suggestion that the levels of nitric oxide being
produced by engines is presently greater than the value for maximum
reactivity rate(9) and to reduce the concentration might only produce a more severe situation than presently exists.   This latter
viewpoint seems to have many critics, including this author.

Because of the thermodynamics involved(22) the oxides of nitrogen
produced at the engine exhaust port are almost entirely nitric oxide
(NO).   This is confirmed by measurements made with mass spectrometer.
(23).   In the presence of oxygen and at low temperature the nitric
oxide becomes nitrogen dioxide (NO_2).   This oxidation usually takes
place in the atmosphere, but it can also commence in the exhaust
system, particularly if air has been added to the gases at the exhaust valve in an effort to partly oxidize the unburned hydrocarbon.
(24)

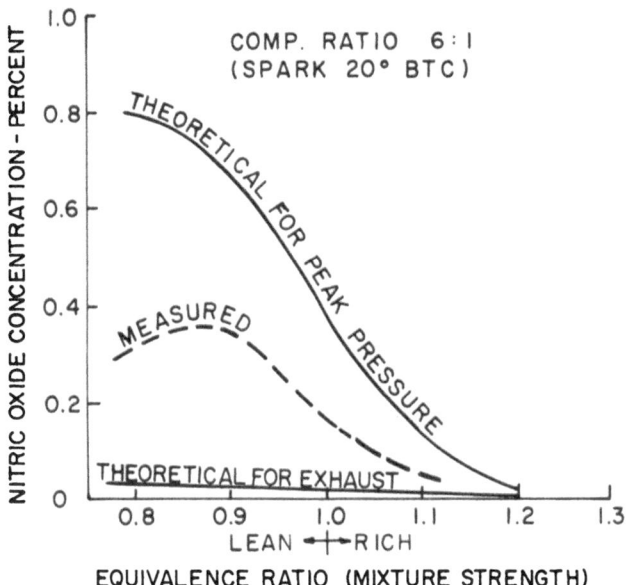

Figure 8. Theoretical and Measured Nitric Oxide Concentrations in
a Spark Ignition Engine.

Nitric oxide is produced in the combustion process in quantities
that are much greater than equilibrium by the same type of mechanism
as for carbon monoxide. Figure 8 illustrates that the amounts are
greater than they should be if equilibrium were the guiding criteria.

It is generally thought that the controlling reaction for forming
nitric oxide is:

$$1/2 N_2 + 1/2 O_2 \rightleftharpoons NO$$

This reaction has an equilibrium constant of 1.37×10^{-5} at $2500°K$
and 2.85×10^{-17} at $298.16°K$.[25] Thus, considerable quantities of
NO can be expected to form in the combustion process, but these
should be reduced to almost zero at exhaust temperatures. The quantities actually measured, as shown in Figure 8, indicate that the
nitric oxide freezes at some relatively high temperature even more
effectively than carbon monoxide.

Many engine variables influence the concentration of nitric oxide.
Figure 9, taken from reference (23), shows the importance of spark
timing.

Compression ratio has a very significant influence on the amounts

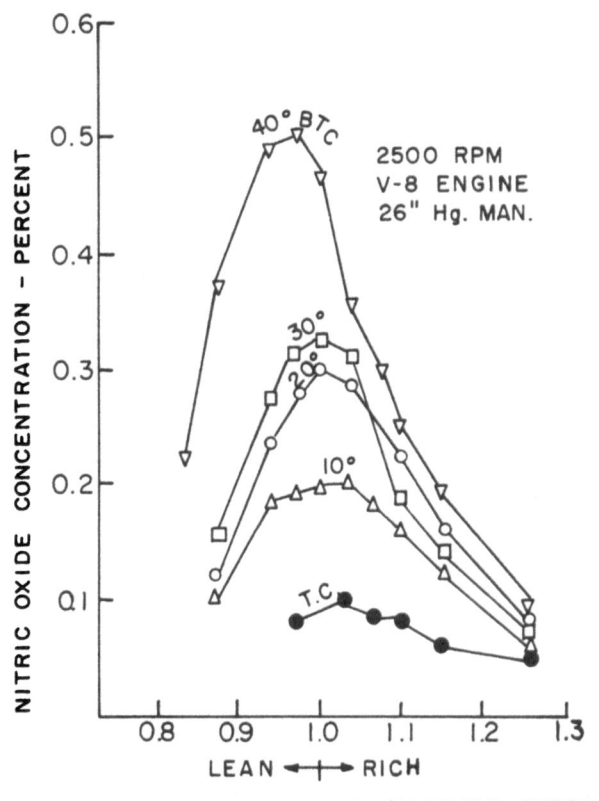

Figure 9. Effect of Spark Timing and Mixture Ratio on Production
of Nitric Oxide from a Spark Ignition Engine.

of nitric oxide produced. Figure 10 from reference (26) illustrates
the typical influence of compression ratio.

    Two things are clear from the knowledge available at this date:
(1) the engine produces only nitric oxide and (2) the reactive
material in the photocatalysis is nitrogen dioxide. Very little
oxide of nitrogen is other than NO or NO_2.

    The Diesel engine produces greater quantities of nitric oxide than
the gasoline engine because of the much leaner mixtures involved.
The gas turbine, on the other hand, produces extremely low concentrations of oxides of nitrogen in its present configuration, as
shown in Figure 11, taken from reference (21).

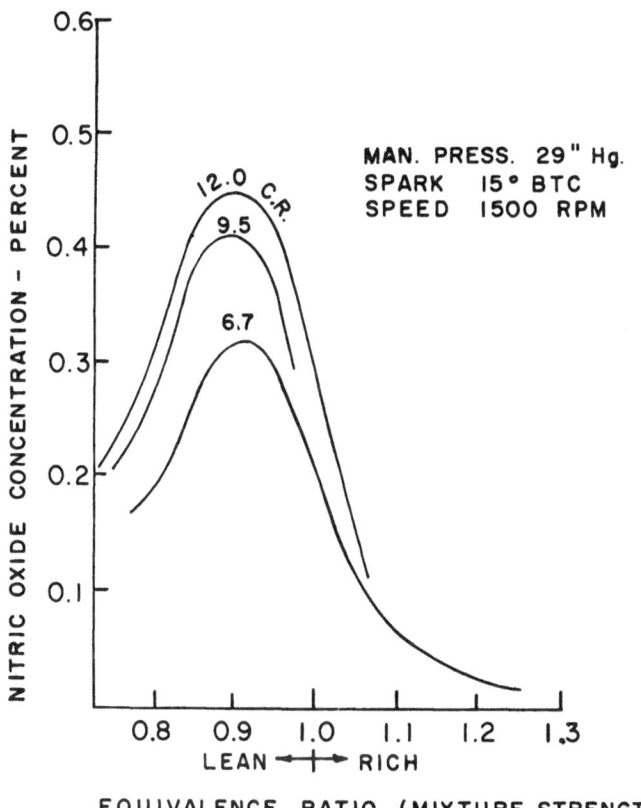

Figure 10. Influence of Compression Ratio and Mixture Ratio on
Production of Nitric Oxide from a Spark Ignition Engine.

D. Carcinogens

The production of carcinogens by engine combustion is also an ar-
gumentative subject. Certain polynuclear aromatic hydrocarbons which
are carcinogenic, such as benzopyrene and benzathracene are known
to be possible of formation in the engine combustion process.(27)
Furthermore, collection of long term high volume air samples in cities with high automobile traffic densities has yielded significant
quantities of these suspected carcinogens.(28)

Reduction in carcinogen concentrations should occur as a consequence of methods used to reduce other unburned hydrocarbons. The
attack on emission of unburned hydrocarbon is also thereby an attack

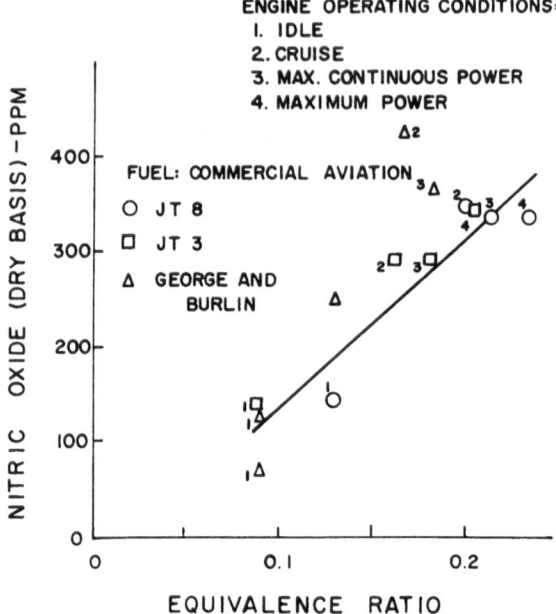

Figure 11. Total Oxides of Nitrogen in the Exhaust of Commercial
Turbojet Engines.

on carcinogens and the positive results should apply equally to both.

There is some reason to believe that Diesel engines may be more
disposed to produce carcinogens than spark ignition.

E. Particulate Matter

Solid particles usually are carbon but also may consist of the
products of lead alkyl decomposition. These particles are undesirable if for no other reason than for the atmospheric occlusion and
surface contamination and discoloration which they produce. There
is good reason to suspect that the surfaces of these particles can
act as sites for chemical reactions. There is also theoretical basis to assume that the particles act as nuclei for the condensation
of vaporous materials into droplets. These droplets then contribute
to atmospheric occlusion in the form of aerosol.

The Diesel engine[29] and aircraft gas turbines are the principal
producers of carbon particles, but the spark ignition engines can
also emit carbon in substantial quantities, such as at full throttle
or when operating at very rich mixture.

F. Lead

Lead alkyls and their accompanying halogenic scavengers are just coming under attack for reasons other than those listed under Parti-culate Matter, outlined above. There is some contention that the concentration of lead in our surroundings is increasing and some of this increase is due to the use of lead in motor gasoline.(30) Counter arguments, both to concentration and resulting influence on health, are being offered. The extent and severity of lead as a problem is not yet clear, but its use is very definitely under attack.

G. Odor

The Diesel engine is the principal offender in the area of odor. Measurements are highly subjective. The materials which constitute the greatest sources of odor seem to be aldehydes.(31)

Obviously any successful attack to reduce or eliminate unburned hydrocarbons will similarly affect exhaust odor.

H. Oxides of Sulfur

Sulfur exits from the engine principally in the form sulfur di-oxide or (SO_2).(32) Its concentration in the exhaust is in direct proportion to the sulfur content of the fuel (excepting for that small amount which combines with and exits with the lead). The re-sults of sulfur dioxide emission are discomfort, because of bronchial and nasal irritation, and visibility reduction because of aerosol formation.(33)

III. EMISSION CONTROL

There is no longer question whether pollution from automobiles can be limited. Experience of the past decade has shown that control is possible. It remains only to decide to what relative degree mo-tor vehicles contribute to pollution and shall be allowed to emit pollutants to the atmosphere. Partly this will be determined by consideration of cost and benefit.

In some communities the problem of automotive contribution to a polluted atmosphere justifies major expenditure. This is particu-larly so if there are problems of health, in addition to physical comfort. Los Angeles is of course a prime example of a community with demonstrated difficulties.

As a consequence, the State of California has generally taken the lead in assessing and attacking problems of air pollution and in acting to harness emissions from the automobile and its internal

combustion engine. More recently the Federal government has accepted
an increasingly larger role in this area. Principally this has been
through activities of the Department of Health, Education and Welfare. However, other branches of the administration, as for example,
the Departments of Transportation and of Commerce have also taken
part. A most notable result of such action is "The Automobile and
Air Pollution"(34) referred to more usually as the "Morse Report"
(for the chairman of the committee which developed it). This publication presents a comprehensive assessment of the motor vehicle
as a contributor to air pollution. Subject matter ranges from the
character and quantity of materials which automobiles generate to
the prospects for alternative power plants. The material to be presented in the following depends heavily for information upon the
same sources as "The Automobile and Air Pollution" and does not differ from it in observations, conclusions or recommendations.

A. Identity of Vehicular Contributed Pollutants

Air pollutants from automobiles can originate at the exhaust, the
crankcase vent, the carburetor or the fuel tank. The more plenti-
ful, complex and difficult to control are the materials produced by
the engine and which issue from the crankcase and the exhaust. They
are potentially also the most harmful. In a vehicle not otherwise
modified for emission control the fuel tank and carburetor will only
contribute approximately 15 percent of the total, (35) and this is
entirely hydrocarbons in an unmodified state.

The exhaust (and engine crankcase, if not connected back to the
intake manifold) can contain varying amounts of all the materials
listed in section II.

Carbon dioxide could also have been added to this list, mostly
because of the so-called "green house effect," but this at the moment shall be ruled outside of the scope of present considerations.

No general agreement exists on the relative damage which each of
the undesirable components of exhaust gas exerts on man, animals and
vegetation. It is quite clear however that most, if not all, can in
a measure be annoying, if not injurious.

The list of vehicle-emitted pollutants easily can be divided into
two groups. The first would include those items which are not a
necessary consequence to engine combustion. As examples, lead and
sulfur could be eliminated from fuel, given sufficient incentive to
do so. Smoke, which is mostly a diesel engine problem, can be controlled by lowering maximum engine power.

The second list consists of the products of combustion from any
kind of burning process. Control of that combustion is the principal

method for varying the relative amounts which will result.

The extent to which the motor vehicles in the United States contributed to air pollution prior to establishment of a nation-wide control (1966) is illustrated by the estimate in Table I. (36)

Table I

Total Estimated U.S. Air Pollution by Source
1966

Source	Tons/Year	Percent of Total
Refuse Disposal	5,000,000	3.5
Space Heating	8,000,000	5.6
Power Plants	20,000,000	14.1
Industry	23,000,000	16.2
Motor Vehicles	86,000,000	60.6

Recognizing that these are total rather than regional figures, it is nevertheless clear that over 60 percent of all pollution came from urban and metropolitan transport of people and goods. The automobile is thus a prime target for attack in the efforts to limit the burden which our form of civilization and culture are placing on the atmosphere.

The extent to which each of the materials emitted by vehicles contributes to this 60 percent of the total is illustrated in Table II.(36)

Table II

Pollutants Discharged by Motor Vehicles

Pollutants	Tons/Year
Carbon Monoxide	66,000,000
Unburned Fuel (Hydrocarbons and Oxygenates)	12,000,000
Oxides of Nitrogen	6,000,000
Sulfur Oxides	1,000,000
Particulates	1,000,000
Lead Compounds as Lead	190,000

Note that the combustion controlled fraction has now been characterized as consisting of three components, carbon monoxide, hydrocarbons (including carcinogens) and oxides of nitrogen.

B. Engine Operating Modes and Severity of Emissions

A great amount of effort presently directed toward the reduction
of emissions makes use of a compilation of typical passenger car
emissions. Table III resulted from the sampling of a very large
number of passenger automobiles undergoing an arbitrary "typical"
Los Angeles driving schedule. (37)

Table III

Typical Exhaust Gas Compositions

Mode of Operation	Unburned Hydro-carbons PPM	Carbon Monoxide Vol. Percent	Nitrogen Oxides PPM	Hydrogen Vol. Percent	Carbon Dioxide Vol. Percent	Water Vol. Percent
Idle	750	5.2	30	1.7	9.5	13.0
Cruise	300	0.8	1500	0.2	12.5	13.1
Acceleration	400	5.2	3000	1.2	10.2	13.1
Deceleration	4000	4.2	60	1.7	9.5	13.0

Legal measures in California and specifications for emission control devices have been based on this kind of information. It is obvious that these numbers are only representative of the particular
driving cycle used and the automobiles involved, but the numbers
give a starting point for attacking the problems of vehicle emissions.

Table III shows that the idle and deceleration modes are the most
productive of unburned hydrocarbons; that idle, acceleration and deceleration produce the highest concentrations of carbon monoxide;
and that cruise and acceleration yield the most oxides of nitrogen.

C. Pollutant Concentration in Exhaust

1. Carbon Monoxide

The propensity for carbon monoxide to effect fatality in humans
and animals has long been recognized. More recently there has been
indication that extensive exposure to far less than lethal quantities
can adversely influence health. Little is yet known about the cumulative biological effects of long time exposure. There is growing

universal agreement that, within reason, all possible steps should
be taken to keep carbon monoxide emissions to minimum levels. It
is obvious from Tables I and II that the automobile engine is a promising point of attack. Fortunately, the gasoline engine has lent
itself to improvement in carbon monoxide emissions.

The reasons for this are embodied in the characteristics which
engines exhibit for production of carbon monoxide. This is shown
in Figure 12. (38)

It can be seen that carbon monoxide appears in the exhaust of
gasoline engines in proportion to the quantity of the fuel in the
fuel-air mixture. Rich mixtures, which give best vehicle performances,
unfortunately also produce the most carbon monoxide. The typical
U.S. car, before the advent of exhaust gas emission control, yielded
an average of almost 4 percent carbon monoxide in the exhaust.

As Figure 12 shows, the simplest way to reduce the quantity of
this undesirable product of combustion is to make the fuel and air
mixtures less rich. The reduction from an average of 10 percent to
2 percent rich has reduced carbon monoxide from 4 percent to 1-1/2
percent.

There are a number of technical problems which accompany such a

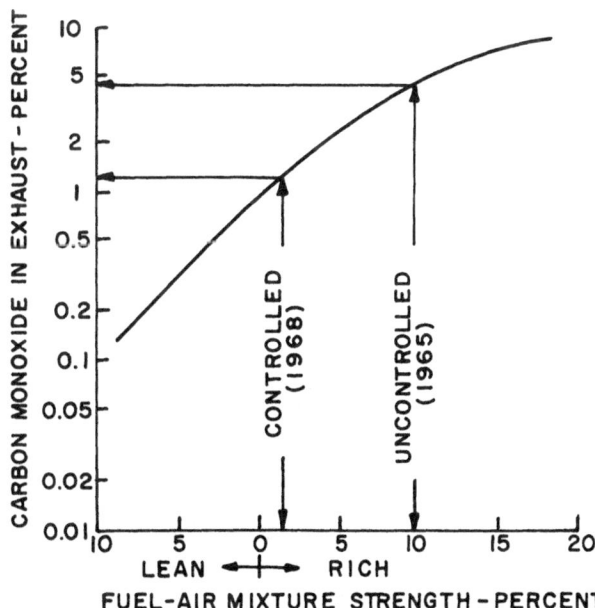

Figure 12.   Influence of Fuel-Air Mixture Strength on Carbon Mono-
             xide Emissions From a Gasoline Engine.

change, however, including what is called drivability. More specifically stated, the engine just does not respond as smoothly at very
lean mixtures and the driver becomes aware that the vehicle is not
operating smoothly.

Nevertheless, 1968 and 1969 model engines have been so developed
that they are operating satisfactorily under conditions which produce an average of less than 1.5 percent carbon monoxide. This improvement has resulted principally from adjustments to the carburetor which cause the engine to run more lean.

2. Unburned Fuel

Lumped into this category are all of those chemicals (except carbon monoxide) which are a consequence of incomplete combustion of
the fuel. These include the hydrocarbons, the partly oxidized fuel
and the polymerized material. The latter, although very low in
concentration, contains the carcinogens. (27)

The quantity and the composition of unburned material vented from
the exhaust is highly variable. Almost every design factor and operating condition has an influence. Nevertheless, certain general observations can be made which relate unburned fuel to engine operation. When considered in a similar manner as for carbon monoxide
the results appear as in Figure 13. (38)

It can be seen that decreasing the fuel-air ratio has influenced
unburned hydrocarbon in a very similar manner to carbon monoxide.
Operation at leaner mixtures has decreased the quantity as measured
in the exhaust. The small disparity in trend between carbon monoxide and unburned fuel at very lean mixture, where unburned fuel
seems to increase, is not believed to be of importance unless future
engines are to be run at very much leaner mixture ratios than can
presently be anticipated.

In considering Figures 12 and 13 it becomes obvious that lowering
of emissions was obtained by a modification in fuel-air mixture in
the direction of less fuel. Corresponding to the previously considered reduction of carbon monoxide is a decline in unburned fuel
from an average of about 1000 parts per million of exhaust (0.1 per
cent volume) to 275 parts per million (0.0275 percent).

3. Oxides of Nitrogen

Customarily the nitrogen oxides produced by an engine are referred to as NO_x. This is because the chemical form could possibly
be either nitrous oxide (N_2O), nitric oxide (NO), nitrogen dioxide
(NO_2) or nitrogen pentoxide (N_2O_5). Experiment shows that the engine produces almost entirely nitric oxide. (23) But this combines
with oxygen in the atmosphere to form nitrogen dioxide. The latter

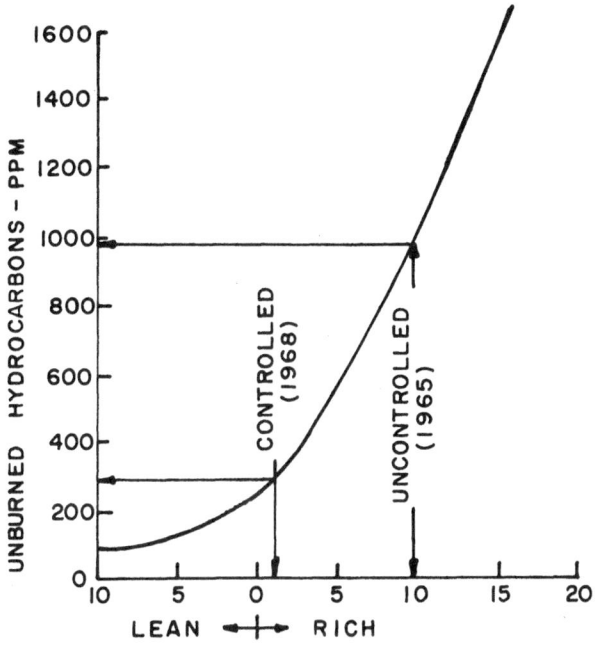

Figure 13.   Effect of Fuel - Air Mixture Strength on Gasoline Engine
Exhaust Gas Hydrocarbon Content

is the important form which is by itself a pollutant, or when photosynthesized with unburned fuel, is the principal component of "smog."
(4)

Again, as with other combustion generated pollutants, engine operating conditions control the quantity of nitric oxide in the exhaust.
Figure 14 illustrates a typical set of results, plotted on the same
basis as for carbon monoxide and unburned fuel. (38)

Illustrated by Figure 14 is an opposite trend with respect to
generation of pollutants than that shown in Figures 12 and 13. Except for extremely low values of fuel-air ratios, nitric oxide production is increased as fuel-air mixture is made more lean. Thus,
a reduction in the amount of fuel supplied to the engines, for the
purpose of lessening emission of other pollutants, results in an increase in oxides of nitrogen.

Figures 12, 13 and 14 have been combined into Figure 15 to illustrate how changes in fuel-air mixture strength can influence the
concurrent emission of the three principal pollutant components of
spark ignition gasoline engines.

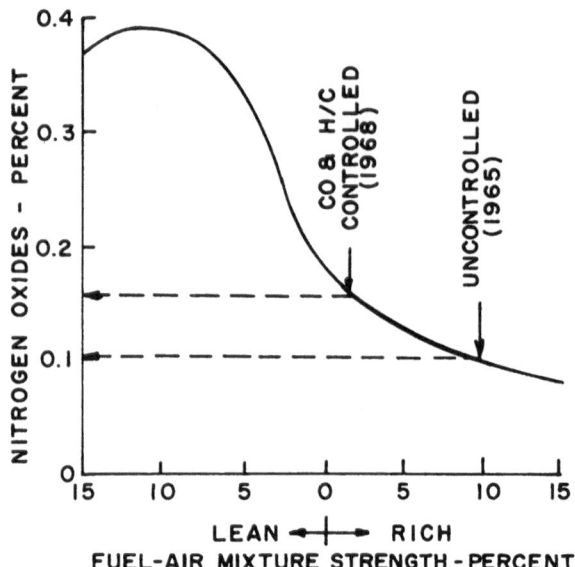

Figure 14.   Influence of Fuel - Air Mixture Strength on Nitrogen
            Oxides Produced by a Gasoline Engine.

D.   Future Control of Emissions

    The obvious question in any discussion of vehicular contribution
to air pollution is that of the lowest level to which the emissions
may be controlled. In the best opinion of those who are involved
in research on this problem, in a six year period it should be technically feasible to produce gasoline engine powered cars with exhaust content of no more than 0.5 percent carbon monoxide, 50 parts
per million of unburned hydrocarbon and 250 parts per million of
nitrogen oxides. (34)

    The techniques to be used in further harnessing the pollution
from vehicles are not yet out of the research and development laboratory. It is known that the application of control systems to effect
the lower emission levels will undoubtedly be more complex and costly
than those used to date and contemplated for the near future. The
cost will in all likelihood be passed on to the consumer. Furthermore, there is the prospect of necessary increased attention to maintenance, not only of the engine itself, but also to the pollution
control systems attached to it and to the vehicle.

E.   Progressive Effect of Vehicular Emission Control

    It can be assumed from experience to date that used cars, even in

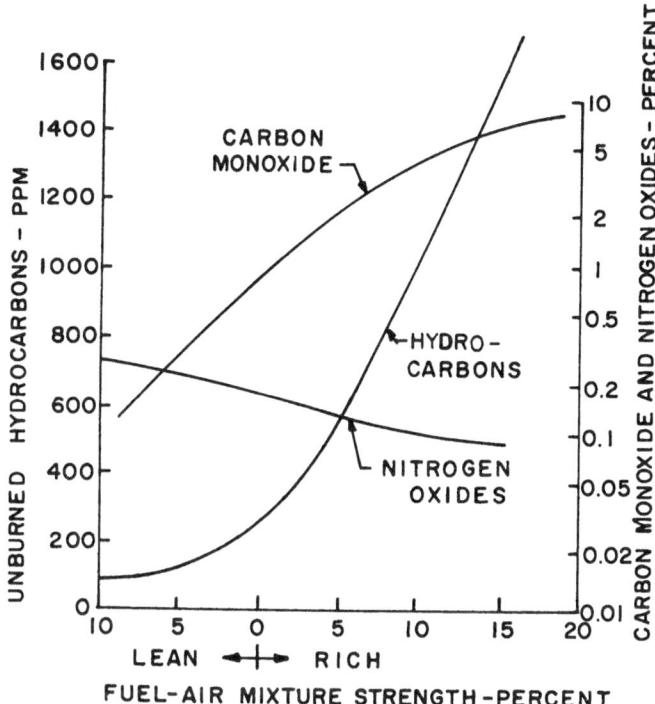

Figure 15. The Combined Effect of Fuel - Air Mixture Strength on the Principal Pollutants in Gasoline Engine Exhaust.

California, will in general not be retrofitted with control systems. Any effect of motor vehicles on the general level of air pollution thus appears completely to be a function of motor vehicle replacement rate.

Figure 16 has been prepared from data in Reference 1 to show how vehicles are phased out of the total car population. What Figure 16 illustrates is the 5 years must pass before half of the motor vehicles on the road are equipped with any device or system introduced as a 100% equipment item. Another factor shown is the significant residue of 10 year old cars on the road.

The implication of such information as contained in Figure 16 is that the problems of air pollution from motor vehicles must of a necessity be anticipated, if future adverse effects on the atmosphere are to be avoided. Further, any control effected immediately will have little influence for a number of years. This is best illustrated by the situation in the Los Angeles metropolitan area.

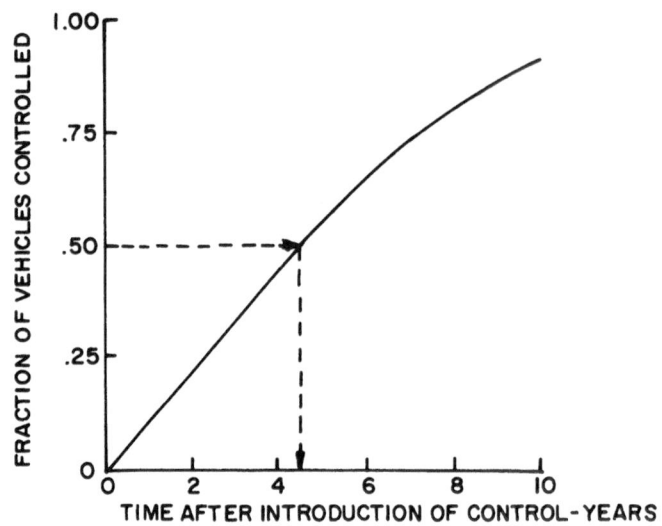

Figure 16. Motor Vehicle Population Distribution

Starting first in California, but now nation-wide, corrective measures have already been taken to limit the extent to which gasoline powered vehicles will pollute the air. Nation-wide, the law now prescribes that carbon monoxide on the average shall not exceed 1.5 percent of the exhaust, and unburned hydrocarbon fuel shall not be greater than 275 parts per million. (39) Nitrogen oxides are not as yet subject to control, although such control is in prospect, at least in the State of California. (41)

The effect that these controls have had on the hydrocarbon air burden showed that the requirements for motor vehicles would only return the quality of the air in Los Angeles to its 1955 level. Furthermore, this was only going to be a temporary low in the curve and would not take place until 1977, a decade after exhaust controls became effective on new cars. Obviously, these existing limitations on emissions had to be supplemented, both by more stringent requirements and by other preventative methods. Table IV indicates how the future limitations appear.

Additionally, a restraint on evaporation from fuel tank and carburetor will also be applied to new cars sold in California, starting with 1970 models. (41) It becomes clear that very stringent limitations on gasoline-powered vehicle emissions must be effected if a city such as Los Angeles is to enjoy the air quality even as it had in 1940.

Table IV

Passenger Car Exhaust Emission Limits (Ref. 40, 41)

Model Year	1968	Federal 1970*	1970	California 1971	1972	1974
Carbon Monoxide, %	1.5	0.82-2.55	1.0	1.0	1.0	1.0
Hydrocarbons, ppm	275	147-457	180	180	125	125
Nitrogen Oxides, ppm	-	-	-	1075	800	350

*According to vehicle engine size and transmission

Thus far the automobile industry has been able to build cars to the emission standards prescribed by law. There is indication that they expect to effect the further improvements prescribed by the limitations imposed in Table IV.

Figure 17 shows the history and future projection of vehicular contributed atmospheric pollution, starting with 1940. (35) Implicit of course is a population growth factor. Los Angeles has been used here as an example because the contribution of motor vehicles to air pollution was early recognized and extensive data taken to chart the changes which have taken place. Equal analysis can be made of any metropolitan or urban area, or air basin.

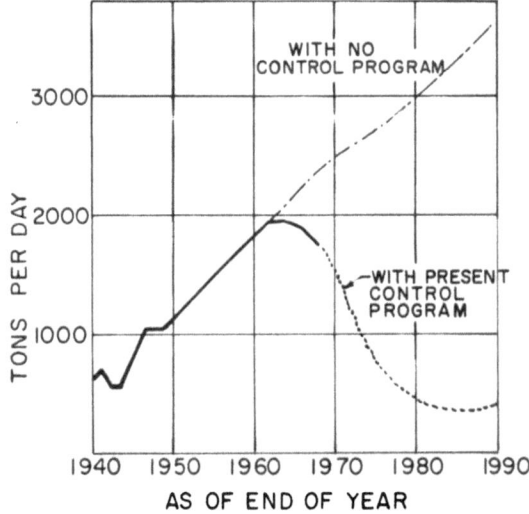

Figure 17.   Emissions of Hydrocarbons from Motor Vehicles in Los Los Angeles County

The principles are the same, whether the effect being studied is due to carbon monoxide, unburned fuel, or oxides of nitrogen.

IV. ALTERNATIVES TO THE PISTON ENGINE

There has been considerable publicity recently with respect to the potential for either electric, steam or gas turbine propulsion of passenger cars. All of these systems promise radical reductions in pollutant generation. Unfortunately, little of this promise is developing as reality. Extensive studies, carried out in connection with Reference 34, indicate why this is so.

Coincident with recognition that the motor vehicle was making major contribution to air pollution was the suggestion that some other power plant might be substituted for the internal combustion engine. Prominent in this burst of enthusiasm was the prospect that electricity might serve the purpose, particularly in the form of batteries or of fuel cells. A fantastic number of technical papers has resulted. In fact it is almost impossible to say anything on the subject which has not already been printed.

This presentation draws heavily from Reference 34 and is for the purpose of developing in technical terms the comparative performance and economy of power plants which could be considered competitive to piston engines.

A. Propulsion Criteria

1. Specific Power

This is the capability to provide a given rate of energy release. Applied to a vehicle, it determines the performance in terms of acceleration, hill climbing ability and maximum speed.

Normally electric power is measured in watts or kilowatts. Thus for comparison purposes power may be related as watts or kilowatts per unit weight. Since the specific power is the measure of the rate at which work can be done, it is exactly the counterpart of horsepower. The relationship is 1.34 horsepower equals one kilowatt, a number very close to 4/3.

2. Specific Energy

This is a determination of capacity, or the total amount of energy capable of being supplied before complete discharge or exhaustion of fuel. Normally the energy stored in a battery is measured and expressed in terms of watt hours, or as specific energy in watt hours per pound. This corresponds to total work and in vehicular application determines the range.

B. Batteries

A representation of the two important characteristics, specific power and specific energy of batteries is shown in Table V, normalized to lead-acid as the most readily available (and cheapest) system.

Table V

Relative Performance of Batteries

Available	Relative Power (Performance)	Relative Energy (Range)
Lead-Acid	1	1
Nickel-Iron	1	1.1
Nickel-Cadmium	1.3	1.3
Silver-Cadmium	2.9	3
Silver-Zinc	4.5	4
Experimental		
Zinc-Air	6	7
Sodium-Air	18	18
Lithium-Copper	9	15
Sodium-Sulfur	5	15
Lithium-Chlorine	7	25

Table V makes it immediately obvious why development effort is being expended on the battery systems in the category of experimental.

It is possible to relate specific power and specific energy for any battery system. This has been done for some of the "available" battery systems listed in Table V. The relationship appears as Figure 18. It is observed that high power demands (performance) results in low energy availability (range).

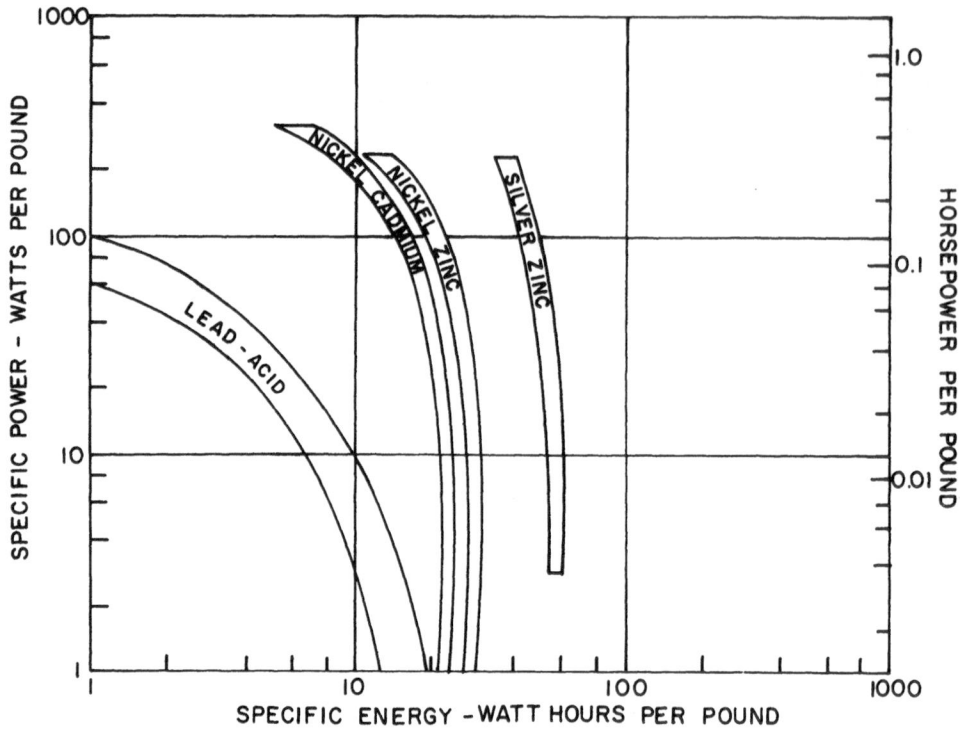

Figure 18.   Comparative Performance of Conventional Batteries

C.   Other Power Plants

The coordinates of Figure 18 are universal. They may be used to characterize any proposed power plant system. As examples, gasoline and Diesel engines, steam or other external combustion engines, gas turbines, and fuel cells can be rated in terms of specific power and specific energy and also plotted. This has been done in Figure 19.

Figure 19 expresses clearly the competition which batteries face as power sources for vehicles. The best system is one which has characteristics which are up and to the right.

Even the best available battery systems fail by a significant amount to match in specific power, vehicular internal combustion piston engines or gas turbines. They are even less attractive from the standpoint of specific energy.

Batteries under development show promise in improved specific energy, but are still not potentially equal in this respect to internal combustion engines.

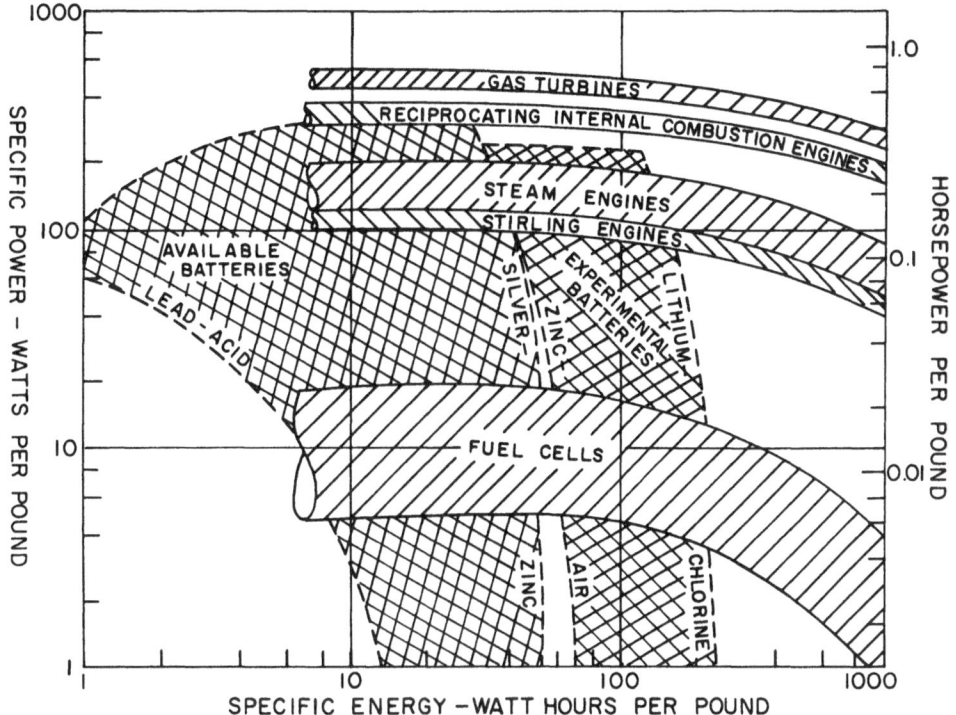

Figure 19. Comparative Performance of Various Power Systems

It can also be seen on Figure 19 that external combustion engines, such as steam and Stirling, as examples, show more favorable specific energy, but lesser specific power than batteries.

Fuel cells do not appear attractive from the standpoint of specific power but do have desirable energy characteristics.

Of even more importance than what is pointed out above is the observation that no known alternative power source can come close to matching the internal combustion engine in the implications of power and energy, which are, of course, performance and range.

D. The Immediate Future

Electric cars, at least for the near future, must be restricted to battery propulsion. Fuel cells, hybrids of various types and "trolley" vehicles are just not within perhaps 10 to 20 years of feasibility. The batteries to be of reasonable first cost, must at present be limited to lead-acid. Maximum speeds are no greater than

25 miles per hour for 40 miles range and 40 miles per hour for 25 miles range. Stop and go considerations hamper these figures even further. Delivery vans and perhaps small urban vehicles constitute the only forseeable reasonable applications. Even so, no presently conceived battery or other electrical propulsion system can come within 50 percent of matching the present gasoline engine in performance, range, or economy.

The steam engine or other external combustion engine comes much closer to satisfying the performance, range and economy criteria which are set by the gasoline engine. Evidence is not conclusive, but emissions from experimental power plants of this type appear to be low. One of the more significant developments in this area is the existing commitment on the part of the State of California to acquire and test highway patrol cruisers fitted with steam power plants. Should this effort be successful, similar engines would by law have to be applied to 25 percent of all vehicles purchased by the State.

Lest too much enthusiasm be generated for the prospect of steam propulsion, it should be stated that the development just considered is still highly experimental in nature. At best, the steam engine will not be able to match its gasoline counterpart in almost any respect except perhaps in lower emissions levels.

Very soon the gas turbine will be in application as almost the sole power plant for aircraft propulsion. It is also in the last stages of development for immediate application to heavy duty ground vehicles. The prospect for use in passenger cars is very high.

A limited amount of data on gas turbine engines indicate that emission of carbon monoxide, unburned hydrocarbon and oxides of nitrogen is below that projected for the gasoline engine of 10 years hence. On a basis to give direct comparison to gasoline engines, automotive type gas turbines produce less than 0.3 percent carbon monoxide, less than 100 parts per million unburned fuel and less than 350 parts per million of nitrogen oxides. (21) In the case of recently constructed automobile gas turbines neither smoke nor odor are detectable in the exhaust.

The principal now holding back the application of gas turbines to passenger cars is cost of manufacture. It may be some time before this obstacle can be overcome satisfactorily.

REFERENCES

1. Automotive Industries, Vol. 140, No. 6, March 15, 1969.
2. Combustion Engine Processes, Lichty, McGraw-Hill, 1967.

3. *Internal Combustion Engines,* Obert, International Textbook Co., Third Edition, 1968.

4. Haagen-Smit, A. J., "Chemistry and Physiology of Los Angeles Smog," *Ind. Eng. Chem.* 44, 1342, 1952.

5. Sweeney, M. P., "Standardized Testing of Smog Control Devices for Vehicular Exhaust," A. I. Ch. E. Meeting, Chicago, Illinois, December 1962.

6. Altshuller, A. P., "Reactivity of Organic Substances in Atmospheric Photooxidation Reactions," A. P. C. A. Meeting, Houston, Texas, June 1964.

7. Hurn, R. W., Dimitriadis, B., and Fleming R. D., "Effect of Hydrocarbon Type on Reactivity of Exhaust Gases," SAE Mid-Year Meeting, Chicago, Illinois, Paper No. 65024, May 1965.

8. McReynolds, L. A., Alquist, H. E., and Wimmer, D. B., "Hydrocarbon Emissions and Reactivity as Functions of Fuel and Engine Variables," SAE Mid-Year Meeting, Chicago, Illinois, Paper No. 65025, May 1965.

9. Caplan, J. D., "Smog Chemistry Points the Way to Rational Vehicle Emission Control," SAE International West Coast Meeting, Vancouver, B. C., Paper No. 650641, August 1965.

10. Hurn, R. W., Fleming, R. D., and Dimitriadis, B., "Influence of Fuel Composition and Volatility on Exhaust Emission and Reactivity," SAE Automotive Engineering Congress, Detroit, Michigan, Paper No. 660113, January 1966.

11. Gagliardi, J. C., "The Effect of Fuel Anti-knock Compounds and Deposits on Exhaust Emissions," SAE Automotive Engineering Congress, Detroit, Michigan, Paper No. 670128, January 1967.

12. Oberdorfer, P. E., "The Determination of Aldehydes in Automobile Exhaust Gas," SAE Automotive Engineering Congress, Detroit, Michigan, Paper No. 670123, January 1967.

13. Papa, L. J., "Gas Chromatography - Measuring Exhaust Hydrocarbons Down to Parts per Billion," SAE Mid-Year Meeting, Chicago, Illinois, Paper No. 760494, May 1967.

14. Maga, J. A. and Hass, G. C., "The Development of Motor Vehicle Exhaust Emissions Standards in California," APCA Meeting, Cincinnati, Ohio, May 1960.

15. Daniel, W. A., "Flame Quenching at the Wall of an Internal Combustion Engine," Sixth Symposium (International) on Combustion, Rheinhold Publishing Company, p. 886, 1956.

16. a. *Vehicle Emissions, SAE Technical Progress Series,* Vol. 6, 1964.

    b. Vehicle Emissions, Part II, *SAE Technical Progress Series,* Vol. 12, 1967.

17. Bartholomew, E., "Potentialities of Emission Reduction by Design of Induction Systems," SAE Automotive Engineering Congress, Detroit, Michigan, Paper No. 660109, January 1966.

18. Cook, F. W., "Antismog Carburetor Hardware and Test Equipment," SAE Automotive Engineering Congress, Detroit, Michigan, Paper No. 660110, January 1966.

19.   Scheffler, C. E., "Combustion Chamber Surface Area, A Key to Exhaust Hydrocarbons," SAE Automotive Engineering Congress, Detroit, Michigan, Paper No. 660111, January 1966.

20.   Bolt, J. A., and Boerma, M., "The Influences of Inlet Air Conditions on Carburetor Metering," SAE Automotive Engineering Congress, Detroit, Michigan, Paper No. 660119, January 1966.

21.   Smith, D. S., Sawyer, R. F., and Starkman, E. S., "Oxides of Nitrogen from Gas Turbines," 60th Annual Meeting, Air Pollution Control Association, Cleveland, Ohio, Paper No. 67-125, June 1967.

22.   Starkman, E. S., and Newhall, H. K., "Characteristics of the Expansion of Reactive Gas Mixtures as Occurring in Internal Combustion Engine Cycles," SAE Mid-Year Meeting, Chicago, Illinois, Paper No. 650509, May 1965. (Transactions SAE, 1966).

23.   Campau, R. M., and Neerman, J. C., "Continuous Mass Spectrometric Determination of Nitric Oxide in Automobile Exhaust," SAE Automotive Engineering Congress, Detroit, Michigan, Paper No. 660116, January 1966.

24.   Reid, R. S., Mingle J. G., and Paul, W. M., "Oxides of Nitrogen from Air Added in Exhaust Ports," SAE Automotive Engineering Congress, Detroit, Michigan, Paper No. 660115, January 1966.

25.   Selected Values of Thermodynamic Properties, National Bureau of Standards Circular 500, 1952.

26.   Nebel, G. J., and Jackson, M. V., "Some Factors Effecting the Concentration of Oxides of Nitrogen in Exhaust Gases from Spark Ignition Engines," Journal APCA, 8, No. 3, November 1958.

27.   Begeman, C. R., "Carcinogenic Aromatic Hydrocarbons in Automobile Effluents," SAE Automotive Engineering Congress, Detroit, Michigan, Paper No. 440C, January 1962. Also SAE TPS-6, 1964.

28.   Collucci, J. M., and Begeman, C. R., "The Automotive Contribution to Air-Borne Polynuclear Aromatic Hydrocarbons in Detroit," APCA Journal, 15, 113, 1965.

29.   Durant, J. B., and Eltinge, L., "Fuels, Engines Conditions and Diesel Smoke," SAE Automotive Engineering Congress, Detroit, Michigan, Paper No. 3R, January 1959.

30.   Patterson, C. C., "Contaminated and Natural Lead Environments of Man," Arch. Env. Health, 11, 344, 1965.

31.   Rounds, F. G., and Pearsall, H. W., "Diesel Exhaust Odor," SAE Transactions, 65, 608, 1957.

32.   Cloud, G. H., and Blackwood, A. J., "The Influence of Diesel Fuel Properties on Engine Deposits and Wear," SAE Transactions, p. 408, November 1943.

33.   Renzetti, N. A., and Doyle, G. J., "Photochemical Aerosol Formation in Sulfur Dioxide-Hydrocarbon Systems," Int. J. Air Poll., 2, 327, 1960.

34.   "The Automobile and Air Pollution: A Program for Progress." U. S. Government Printing Office, U. S. Department of Commerce, Report of the Panel on Electrically Powered Vehicles, October 1967.

35.   Maga, J. A., and Kinosian, J. R., "Motor Vehicles Emission Standards--Present and Future." SAE Automotive Engineering Congress, Detroit, Michigan. Paper No. 660104, January 1966.

36. "The Sources of Air Pollution." U.S. Government Printing Office, Department of Health, Education, and Welfare, 1966.

37. Teague, D. M., et al., "Los Angeles Traffic Pattern Survey," SAE National West Coast Meeting, Paper No. 171, August 1957. Also SAE TPS-6, 1964.

38. Starkman, E. S., "Engine Generated Air Pollution--A Study of Source and Severity." FISITA, Munich, June 1966.

39. Public Law 89-262, 89th Congress, S. 306 U.S. Government Printing Office, Washington, D.C. October 20, 1965.

40. "The Cleaner Air Act of 1968", U.S. Government Printing Office, Department of Health, Education and Welfare, 1968.

41. "Pure Air Act of 1968", State of California, Assembly Bill #357.

42. George, R. E., and Berlin, R. M., "Air Pollution from Commercial Jet Aircraft in Los Angeles County," Air Pollution Control District/County of Los Angeles, April 1960.

EFFECTS OF AIR POLLUTION ON HEALTH

Jay A. Nadel

Cardiovascular Research Institute

University of California School of Medicine

San Francisco

Anatomy and physiology of the lungs as a basis of understanding effects of inhaled pollutants.

A. Anatomy: Breathing takes place through either the nose or mouth. The nose contains tortuous channels which allow the inspired air to be exposed to a maximum surface area. These channels warm and moisten the air, and highly soluble substances are prevented from entering the deeper lung structures when inhalation occurs through the nose. The nasopharynx connects the nose with the pharynx which in turn ends in the larynx (or voice box). Below the larynx, the airway continues via a single tube, the trachea. Inside the chest, the trachea subdivides into progressively branching tubes of decreasing size. As a consequence of the increasing number of branches, there is an increase in cross sectional area, so that inflowing air decreases its velocity when it reaches the very small airways. As a consequence of the changing dimensions of the airways, most of the resistance to airflow is normally located in airways whose diameters are approximately 3-7 mm in diameter. The airways are designed to distribute the air as uniformly as possible to the many millions of minute air sacs called alveoli where gas exchange occurs. The airways are designed to minimize the volume of the conducting tubes (which do not participate in gas exchange) but at the same time not so small that the resistance to airflow is excessive (this would increase the work of breathing). The airways are lined by cells called epithelial cells. These cells contain various elements designed to protect the lungs from trauma. These elements vary in different depths of the lungs: the larger airways contain "goblet" cells which secrete mucus, which coats the airways. In disease, these mucus-producing cells sometimes becomes enlarged

239

and produce an abnormal amount of mucus which plugs the airways.
On the tips of the epithelial cells are cilia, fine hair-like structures which sweep rhythmically, pushing material on the surface
of the airways toward the nasopharynx where they can be eliminated;
cilia exist in airways at all levels, and assist the cough mechanism
in cleansing the airways. Some noxious materials paralyze the cilia,
thus allowing the accumulation of mucus and other materials which
deposit on the surface of the airways to accumulate.

The airways contain nerve endings which are close to the surface of the airways and which respond to mechanical or chemical
irritation. Stimulation of these nerve endings results in cough,
a reflex which protects the organs by propelling materials out of
the airways. Cough is efficient in removing materials from only
the large conducting airways where velocity of flow is sufficient;
in small airways, airflow is slow and forces are insufficient to
shear foreign materials from their walls, so only ciliary activity
functions at these levels to remove foreign materials.

The walls of the airways contain smooth muscle which has the
capacity of contracting and thus narrowing the diameter of the airway. Contraction of the smooth muscle increases the resistance to
airflow and increases the work of breathing; in some diseases, this
smooth muscle contraction may be severe (e.g., in asthma) and lead
to marked dysfunction of the respiratory apparatus. The airway
smooth muscle may contract due to direct stimulation or indirectly
via reflexes.

The airways end in minute air sacs called alveoli. Inspired
gas is carried into the alveoli during inspiration where it comes
into close contact with the blood which flows through the pulmonary
circulation; the blood travels into the lungs via two large pulmonary arteries which also subdivide into progressively smaller vessels
until they form minute capillaries which have thin walls and are in
intimate contact with the alveoli. Thus, gas exchange occurs at the
level of the alveolar-capillary membrane, where the distance for
diffusion is small (less than 0.1 micron thick) and O_2 can move into the blood and CO_2 can be removed from the blood. There are about
300 million alveoli, which are so designed to provide a vast and
extremely thin surface for the transfer of gases between air and
blood. The blood tubes are also designed to allow a large area
for gas exchange but at the same time to minimize the resistance to
blood flow.

The alveoli act like bubbles which expand during inspiration
and contract during expiration. They are lined by a thin layer of
a material called "surfactant" which lowers surface tension and
allows the small alveoli to remain open even when the alveoli become
small (e.g., during expiration). When this surfactant is destroyed
or inactivated, the lungs tend to collapse. Certain toxic materials

which reach the alveoli may injure the lungs by interfering with
normal surface activity.

Mobile cells (called "macrophages") engulf and carry away
insoluble particulates which deposit deep in the lung spaces. The
purpose of this process is to prevent these particles from injuring
the alveolar epithelium and to facilitate removal of the particles.
However, certain dusts may be injurious to the phagocytic cells and
ultimately cause their death, and this may interfere with normal
clearance mechanisms.

Effects of air pollutants on lungs: The effects of inhaled
substances on the lungs depends to a large extent on the physico-
chemical characteristics of the material: when gaseous materials
are inhaled, the site of their effects depends greatly on the solubility of the gas. SO_2 is highly soluble and is removed from inspired air in the large conducting airways, so that very little of
the material reaches the distal airways and alveoli. Nitrogen dioxide, on the other hand, is less soluble and reaches the tiny airways and alveoli where it causes its damage. When particulates are
inhaled, the size and other physical characteristics of the particles
determines the site of deposition (and thus the site of primary action).

B. Effects on Airways: Sulfur dioxide is the model of substance which has its major effect on the airways. When SO_2 is inhaled, its concentration decreases rapidly, so that almost no SO_2
penetrates lower than the larynx. In low concentrations (2-5 P.P.M.)
inhalation of SO_2 constricts the lower airways by stimulating receptors in the large airways. These receptors fire electrical signals which are transmitted to the central nervous system and thence
back to the airways smooth muscle via another nerve called the vagus.
This reflex contraction of the airway smooth muscle narrows the airways. In higher concentrations, the airway narrowing becomes more
severe and may hamper ventilation. SO_2 causes airway constriction
in doses which may be too small to be detectable by the subject
(i.e., by taste or smell). The airway constriction can be prevented
by administration of drugs which paralyze the smooth muscle and
prevent it from contracting.

Air pollutants may also narrow the airways by causing swelling
of the membranes lining the airways, by enlargement of the glands
which produce mucus, or by plugging the airways by the mucus after
it has been discharged from the mucus glands, by direct action of
the irritation on the airway smooth muscle. (Various effects of
different concentrations of air pollutants are listed in the tables
of the handbook entitled Environmental Biology. P. E. Altman and
D. S. Dittmer, eds. Feder. Amer. Soc. Exp. Biol., Bethesda, Maryland,
1966, pp 276-281.)

I. Effects on Alveoli and Lung Tissue:

(1) Acute Effects: Materials which reach the alveoli may
injure the lining epithelial cells, the capillaries or may interfere with the alveolar surfactant. As a result of this insult, the
alveoli may collapse or may become filled with "edema" fluid; these
damaged alveoli may not ventilate properly and inspired air may not
reach the appropriate lung capillaries, causing serious physiological
consequences. (Various effects of different concentrations of air
pollutants are listed in the appropriate section of Environmental
Biology. P. E. Altman and D. S. Dittmer, eds. Feder. Amer. Soc.
Exp. Biol., Bethesda, Maryland, 1966, pp. 281-284).

(2) Chronic Effects: Recurrent injury to lung tissue may
result in the production of scar tissue which contracts and also
interferes with proper gas exchange. Some inhaled materials have
no serious acute effects when inhaled, but they are deposited in
lung tissue and remain there (i.e., silica). After many years,
the silica particles stimulate a reaction of the lung tissue, and
progressive scarring of the lungs occurs.

II. Effects of Aerosols:

When materials are inhaled as particles, their effects also
depend on the site of deposition, which in turn depends on the physical characteristics of the aerosol. The deposition of particles
in the respiratory system depends on three physical forces: (1)
Inertial forces cause deposition in the nasopharynx and at points
of branching of airways wherever velocity of flow changes. Inertia
is most important in the respiratory system in the large airways,
especially when rapid forced breathing occurs; its importance decreases with depth in the respiratory system. (2) Gravitational
sedimentation is proportional to particle settling velocity and the
duration of time available for settling; because of the decreased
velocity in the fine airways, gravitational effects are increased.
(3) Diffusion is most important for very fine particles which are
deposited on the walls of the finest airways and within the alveolar
spaces. The magnitude of the diffusional force is significant for
particles above 0.5 micra.

Aerosol particles are important sources of irritant materials,
not only by the nature of the particles but because of their ability
to absorb other materials of their surfaces. Thus, Amdur has shown
in experimental animals that the effects of various inhaled gases on
airway resistance (e.g., SO_2) are potentiated by chemically inert
saline aerosols. Aerosols are also important because they create
a problem not only at the time of arrival on the surface of the lung
or airway, but because they must be cleared from the area to prevent cumulative effects.

III. <u>Clearance Mechanisms</u>:

    1. Airways:

        (a) Ciliary Action: There are approximately 20 cilia per epithelial cell in the airways, and these cilia beat about 1300 times per minute. This activity results in the propulsion of the overlying "mucous blanket" at an average rate of 15 mm. per minute. When the cilia are paralyzed, this clearance mechanism is also paralyzed.

        (b) Mucous Blanket: The fluid film covering the airway is secreted by the underlying cells. Alterations of this mucus will change the ability of the cilia to propel it upward and thus will impair clearance of the materials the mucus carries with it.

        II. Alveoli: Materials which are deposited in the alveoli can be removed either by transport up the airways or through the lung tissue to lymph nodes.

Cigarette smoking and exposure to uranium by industrial workers are causally related to lung cancer, but the relationship of lung cancer to community pollution is debated. Experimentally, lung cancer has been produced in mice by intermittent inhalation exposures to organized gasoline vapors and viruses and in rats exposed to sulfur dioxide and benzpyrene. (Stokinger, H. E. and Coffin, D. L.: Biologic effects of air pollutants in <u>Air Pollution</u>, Vol. 1. A. C. Stern, eds. Academic Press, New York, 1968, pp. 445-546. Further studies are needed in humans.)

IV. <u>Effect of Inhaled Materials in Patients with Lung Diseases</u>:

    1. Increases "reactivity" of Airways to Inhaled Irritants:

        Some patients with obstructive airway diseases (e.g., asthma and bronchitis) react to inhaled irritants such as sulfur dioxide by constricting their airways and by coughing at a dose of the irritants which would be expected to have no effect on healthy individuals. In addition, further narrowing of airways by inhaled irritants in patients who already have constricted airways may have more serious consequence than it would have in healthy individuals.

V. <u>Do Air Pollutants Cause Lung Disease?</u>

It seems probable that sufficient exposure to certain types of air pollution is a causal factor in chronic bronchitis and emphysema, but it is difficult to prove conclusively. However, epidemiologic studies show that it is almost certainly a promoting or aggravating factor. For example, in Great Britain increased mortality and aggravation of symptoms of chronic bronchitis are associated with in-

creased concentrations of smoke, particulates and sulfur dioxide.
In the occasional disastrous air pollution episodes the relationship
between exposure and effect is more clearcut; in these episodes,
the aged and the very young were most affected. Particulates in the
alveoli stimulate the outpouring of alveolar macrophages and pro-
tein-like material into the alveoli, and the macrophages engulf the
particulates. Some particles (e.g., silica) are toxic and cause
death of the macrophages with subsequent laying down of scar tissue in the lungs. This scar tissue has deleterious effects on
respiratory function.

VI.   Protective Effect of Previous Exposure:

     In animals previously exposed to low levels of ozone, subsequent
administration of doses that previously would be lethal do not result in death. The mechanism of this protection is unknown, and
its clinical implications are unclear.

VII.   Systemic Effects of Air Pollutants:

     When materials are inhaled and deposited in the lungs, they
may be absorbed into the blood stream and cause deleterious effects
in other organs. Carbon monoxide is an important example of a pollutant whose major effects occur after absorption: This odorless,
colorless gas has its major origin in the incomplete combustion of
carbonaceous material and has its major effect on the organism because of its strong affinity for hemoglobin (the pigment contained
in red blood cells which is responsible for transporting O_2 to the
tissues).

     Carbon monoxide combines with hemoglobin 210 times as readily
as does O_2; thus, it effectively prevents hemoglobin from performing
its important function of transporting O_2 from the lungs to the tissues. Carbon monoxide is also excreted via the lungs. Significant
levels of carbon monoxide may be derived from cigarette smoke and
from auto exhausts. Since the effects of this pollutant derive
from its impairment of O_2 carrying capacity, it would be of most
serious import-- (a) at high altitude, or (b) in patients with
diseases of the heart, lungs, or central nervous system, or (c)
in the presence of anemia.

VEGETATION DAMAGE FROM AIR POLLUTION

Ellis F. Darley

Statewide Air Pollution Research Center
University of California
Riverside, California

Introduction

Air must be considered as an important natural resource vital
to animals and plants for their life processes and to industry and
transportation for combustion purposes. The quality of the air,
or the chemical composition of its minor constituents, varies as
a result of emission of contaminants from all of man's activities
including the generation of energy, manufacturing of goods, and
disposal of various types of wastes. The amounts of the major
constituents of air--nitrogen and oxygen--are fairly constant and
together account for ninety-nine per cent of the atmosphere near
the earth's surface. In clean air the remaining one per cent is
made up of carbon dioxide, water vapor and a variety of other compounds. The minor gaseous constitutents of the air, however, often
vary as a result of emissions from the various activities noted
above. Thus the quality of the air, when altered by many of these
contaminants, may have profound effects on the growth and development of vegetation because gases enter the leaves during the course
of normal gas exchange and, if toxic, kill cells and tissues and
otherwise interfere with the plant's normal growth processes.
The problem is aggravated as urban communities enlarge and gradually merge with the once distinct surrounding rural areas. Outstanding examples are the Los Angeles basin and the San Francisco
Bay area in California and along the east coast from Washington
to Boston.

The object of this chapter is to discuss the principle air
pollutants that adversely affect growth and development of vegetation.

The Pollutants

The pollutants to be considered fall into two broad classes; (1) gases and (2) particulates.

Gases-- Some of the gases are those related to the photochemical process. Photochemistry and photochemical air pollution have been discussed in some detail in previous chapters. It is only necessary to say here that the photochemical products affecting plant life result from sunlight-initiated reactions between certain hydrocarbons and oxides of nitrogen. The automobile is the primary source of the reacting pollutants but they also come from the combustion of any organic fuels. An important group of photochemical products are the peroxyacyl nitrates, of PAN's, which are very toxic to plants and which were described only as recently as about 1956. The basic formula is $\overset{O}{\overset{\|}{R \cdot C}}-OONO_2$. Peroxyacetyl nitrate, or PAN, has two carbon atoms; peroxypropionyl nitrate, PPN, has three carbons, and peroxybutyrl nitrate, PBN, has four carbons. All three of these PAN's have been made in the laboratory but only the first two have been found in the atmosphere. Where they occur simultaneously, their concentration ratio has been about 10 to 1. This is ironically fortunate because PPN is about 2-3 times more toxic than PAN; PPB is even more toxic.

Two other photochemical products of importance in vegitation damage are ozone and nitrogen dioxide.

Closely associated with photochemical air pollution because of similarity of source, is the hydrocarbon, ethylene. So far as is known, ethylene is the only hydrogen having direct effect on plants at ambient concentrations. Propylene and acetylene most nearly approach the toxic activity of ethylene, however, from 60 to 100 times as much gas is required for comparable effects.

Sulfur dioxide and hydrogen fluoride are the two remaining gases of principle interest. These arise from a variety of stationary industrial sources.

Particulates-- Particulate materials involved in damage to vegetation are sulfuric acid droplets which settle out of the air, and settleable dusts from a variety of industrial operations.

General Considerations

Before going into specific effects of these pollutants on plants, it would be well to review some general considerations involved in the relationship between pollutants and vegetation damage.

Plants are very useful indicators for pollutants and are often the first evidence that an air pollution problem is developing.

The pollutants noted above often produce characteristic marking on given plant species thus providing a means of pollutant identification. Vegetation first indicated a new problem in Los Angeles in 1945 which we now recognize as photochemical air pollution. Vegetation surveys made at regular intervals revealed the gradual spread of pollution in the Los Angeles basin as well as its occurrence in principle population centers of the state. Similarly, vegetation damage proved to be the first evidence that a photochemical pollution problem was developing in Denver in 1953 and along the east coast in 1956 and 1957. Additional surveys by competent observers now indicate that the problem is present in some 31 other states in the union. Damaged vegetation of the same type has been found recently in Holland and Germany.

Plant injury around stationary sources is usually localized, perhaps within a few miles depending upon the terrain and wind patterns. Several such sources even though located in the same air basin may be distinguished from one another through local vegetation surveys. With mobile sources, on the other hand, the affected areas may be quite large and even encompass an entire air basin as occurs in Southern California from Ventura County south to San Diego, and inland beyond San Bernardino and Riverside.

The effects of pollutants on plants are usually noted on the leaves because the leaves are the site of gas exchange and the photosynthetic process. In some instances flowers are effected without noticeable effect on the leaves.

With gases, the quantities of contaminants required to cause injury are very small, usually less than 1 part per million (ppm) and sometimes as low as 1/10 part per billion (ppb). Similarly, the length of exposure required for many pollutants is usually relatively short. For example, a few hours of ozone at .2 ppm or of PAN at .05 ppm will cause rather distinctive injury. A few days or weeks may be required for hydrogen flouride in the ppb range to cause injury.

The severity of injury varies with the pollutant, the concentration, duration of exposure, and plant species. Generally, two types of visable injury may be noted: (1) Acute injury in which markings on the leaves of susceptible plants are quite characteristic and may result in death of cells or tissues. Such injury occurs when the rate of absorption of gas exceeds capacity of tissues to oxidize, reduce, respire, or translocate the pollutant; (2) Chronic injury which is usually not characteristic wherein there is a chlorosis or yellowing of leaf tissue and usually no death of cells. Chlorosis can be confused with the effects of a multitude of other causes such as improper nutrition, water imbalance, and freezing. Chlorosis or chronic injury occurs when

the rate of absorption is slower and the plant is able to accommo-
date the toxicant with less serious effect. A third and very important effect of pollution on plants, but one that is not easily
visible, is suppression of growth. This occurs when the plant is
exposed to a subnecrotic or sublethal concentration. The plant
does not grow as well or yield as much product of good quality.
This effect is often called hidden injury, but the term is inaccu-
rate and should not be used. Growth suppression in the absence of
visible symptoms occurs when the rate of absorption of the gas is
such as to result in alterations in photosynthesis and respiration,
interference with enzyme activity, and changes in cell wall perme-
ability.

The losses in plants vary according to the type of effect, that
is whether acute or chronic injury or suppression of growth, and
also with type of plant grown. In the case of leafy vegetable,
such as spinach or romaine, or with flowers, such as orchids and
carnations, wherein the marketable portion of the plant is injured,
the grower may sustain direct or complete loss. For example, if
a spinach or romaine field is within a week or so of harvest and
an air pollution episode occurs, the marking on the leaves may be
so severe that the product is unsightly and the housewife would
not consider buying them. Furthermore, it may not be profitable
for the grower to sort out the injured leaves of the spinach and
try to sell what was left. Also it might not be feasible to remove the injured leaves from the head of romaine because this would
also make the product unsightly and the costs might be prohibitive.
On the other hand, if the plant involved was a root crop such as
carrots or table beets or one where the fruit is used such as tomatoes and beans, comparable injury to leaves causes an indirect
loss due to decreased yields and impaired quality. Thus several
of the leaves of the affected plant may be killed, but the leaves
are not being sold; it is the root or fruit that is marketed.
Nevertheless, injury to the leaves interferes with the growth
processes of the plant sufficiently so that the yield of the marketable product may be reduced. Very rough estimates of annual
dollar losses from direct effects of air pollution in California
have been assessed at from $10 to $12 million and from indirect
losses at another $100 million.

In contrast to gases which enter into the plant through the
small pores on the lower sides of the leaves, particulates
generally settle on leaf surfaces and form undesirable deposits.
In the case of alfalfa even if there is no direct damage to leaves,
just the presence of the deposit on the leaf will tend to reduce
the market value of the product. In some cases dusts may be somewhat toxic and are responsible for chlorosis and death of leaf
tissues.

Specific Effects of Pollutants

Peroxyacyl nitrates--The typical injury symptom from PAN's is a silvering, glazing, or otherwise metallic sheen on the lower surface of affected leaves with no injury to the upper surface. The injured leaf tends to dry so that with time injury may extend to the upper surface and large portions of the leaf will appear dead or necrotic. The symptoms from various PAN's appear to be the same but as the carbon number of the molecule increases so does the relative toxicity. PPN is at least twice as toxic as PAN. Among the best indicators of the presence of these gases are the petunia, spinach, and lettuce.

In addition to injury symptoms, there are physiological and biochemical effects from concentrations insufficient to cause marking. For example, the rate of water uptake of exposed plants may be reduced as much as 25 per cent. Blooming of petunia and fruit set in tomatoes may be reduced. Also the growth of tomato may be reduced by as much as 40 per cent. Results of experiments just completed show that field grown citrus trees are reduced in growth and fruit production is impaired. PAN interferes with enzyme systems and cell wall metabolism.

One interesting feature about PAN is that light is a requirement for injury. If plants are placed in darkness for relatively short periods either before, during, or following exposure to the pollutant, no injury occurs. These experiments were conducted in such a manner to insure that the stomata, or the pores, on the lower leaf surfaces were open. Thus, some light dependent activity of plant cells ultimately controls whether PAN will have any effect, but the light wave length is different from that which controls photosynthesis.

In addition to causing plant injury the PAN's are also responsible for some of the eye and throat irritation experienced in photochemical air pollution episodes. Of all the air pollutants of which we are aware, PAN is the only one that has an effect both on plants and animals.

Ozone--Typical injury symptoms from ozone are stippling, mottling, or a chlorosis that is confined to the upper surface of the leaf. The color of the marking varies from a light tan to a red to almost black depending upon the species affected. Ozone enters the plant through the same pores on the lower leaf surface as do other pollutants, but for some reason has a greater affinity for the chlorophyll bearing cells near the upper surface of the leaf. It is able to migrate through the leaf and effect the chlorophyll bearing tissue near the upper surface and this is the reason for the effect being present mainly on the upper surface of the leaf.

Thus ozone and PAN, two photochemically produced pollutants, can be distinguished by type of markings and position on the leaf.

Ozone is responsible for serious marking and premature needle drop in the pine forests and Christmas tree plantings on the east and west coasts. Recent surveys indicate that more than 25 thousand acres are affected in the San Bernardino mountains east of Los Angeles. Also several of the important pine forests in the mountains east of the San Joaquin Valley of California are beginning to show signs of ozone injury.

Ozone, as does PAN, has a somewhat similar suppressing effect on growth and cell wall metabolism, but the effect of light is different than with PAN. Whereas short periods of darkness before, during, and after exposure prevents PAN damage, this is not so with ozone. Very long periods of darkness of 48 to 72 hours are required before the severity of injury is decreased and this is related to depletion of reserve carbohydrates in leaf tissues. The sensitivity of such plants can be restored by adding sugar through the cut stems. In other words, the reserve carbohydrates are restored.

Ethylene--Of the several hydrocarbons produced from combustion sources, ethylene is the only one known to have direct effect on plants at ambient concentrations. Ethylene is closely associated with episodes of photochemical air pollution although the gas is not very active photochemically. The half-life of ethylene is about 4 to 6 hours, whereas that of the most reactive hydrocarbon may be as short as 20 minutes. The product of ethylene reaction is formaldehyde; no PAN is formed. The important thing about ethylene is that it does persist in the air for periods of time sufficient to cause damage to a variety of plants.

In the main, ethylene affects flower production of ornamental species. The dry sepal injury to orchids is a prime example. The tips of the sepals become dry and discolored. This makes the flowers completely unmarketable. Some estimates indicate a $75,000 annual loss by just two growers in the San Francisco bay area. During the winter months as much as 80 percent of the daily flower production may be lost. Another important effect is the shatter or the abcision of the colored portion of the flowers of snapdragons. The flowers on the spike fall to the ground leaving a naked stem; of course this makes the flower completely unsaleable. Another effect of ethylene is that which is called sleepiness of carnations. If the flowers of carnations are fully opened and then exposed to ethylene, they will close. Developing flowers fail to open properly. The pollutant may have similar effects on flowers of a variety of field crops but this phase has not been investigated thoroughly. Research is now underway at Riverside to determine the effect ethylene has on leaves and on the growth of a variety of plants. Other hydrocarbons are also being investigated although most of the evidence to date indicates that no other hydrocarbon is harmful at ambient concentrations.

Nitrogen dioxide--It was once thought that this pollutant was only important because of its participation in the photochemical process. Eight to 50 ppm were required to mark plants and these concentrations far exceeded those found in the atmosphere. Acute symptoms from these high concentrations are very similar to those noted below for sulfur dioxide. With sulfur dioxide, however, only about 1 ppm for a few hours is required. Recent experiments with a variety of plants exposed to nitrogen dioxide have demonstrated that concentrations of a few tenths ppm for several weeks could reduce growth as much as 35 per cent. At these very low concentrations there are no acute symptoms on the leaves, but there is a definite demonstration of growth suppression. For reasons as yet unexplained, plants are more susceptible to the effect of NO_2 at night than during daylight hours. Considerable research is now underway both in the U.S. and in Europe to learn more about this important pollutant.

Sulfur dioxide--This is probably the oldest known phytotoxic air pollutant. Investigations have been underway for over one hundred years. Early work on sulfur dioxide resulted from litigation over damaged vegetation in the vicinity of smelter operations in Europe and North America. Examples are the Selby smelter in the San Francisco Bay area around 1915, and the Trail smelter in British Colombia in 1930. The latter smelter caused much damage in the Colombia River Valley in the U.S. necessitating an international commission to settle this problem. The typical symptom of acute injury from sulfur dioxide is a white to tan bleaching of leaf tissues. The injury goes clear through the leaf from surface to surface and is not confined to one or the other surface as is the case with the photochemical air pollutants, PAN and ozone. Thus SO_2 is distinguished from ozone by the nature of the marking and the position on the leaf. There are nutritional imbalances, extremes in temperature, and some organic plant diseases which produce the symptoms similar to SO_2 and care must be exercised in plant surveys to distinguish these various effects.

More than 350 plant species have been rated in relative susceptibility: Alfalfa, barley, cotton and lettuce are examples of the most sensitive plants, while gladiolus, corn, citrus and oak are among the most resistant. When SO_2 is first absorbed by a plant, a sulphite is formed within the tissues, which is then slowly transformed to sulphate. It has been shown by several researchers that sulphate is about 30 times less toxic than sulphite. About all plants are injured by the same level of sulphite in their cells, but they differ widely in their reaction to SO_2 in the atmosphere due to variations in rate of absorption. Chronic injury, a rather non-characteristic chlorosis, results from exposure to lower concentrations over a longer period of time permitting the plant to transform the absorbed sulphite to the less toxic sulphate.

To date there is no clear cut evidence that concentrations which do not mark plants will cause growth suppression. This in contrast to the photochemical products which very clearly do suppress growth at relatively low concentrations. For short term experiments with alfalfa it was shown that reduction in yield was directly related to the amount of tissue destroyed by SO_2. Results were duplicated by mechanical clipping of equivalent amounts of leaf tissue. Failure to demonstrate growth suppression may be due to insufficient length of exposure.

Recent and very important research has shown that there is a synergistic reaction between low, non-lethal concentrations of SO_2 and ozone. That is, low levels of either gas which cause no symptoms when acting alone, do cause symptoms when acting together and the symptoms resemble ozone. These results have very important implications for eventual air quality standards of these two gases.

Because of satisfactory control in many areas, acute injury seldom occurs in California. However, accidental spilling occasionally happens as occurred recently near a refinery in Los Angeles. Ten to 20 ppm for a few minutes caused serious injury to nursery and surrounding field crops.

Hydrogen fluoride--The typical symptom is a necrosis of the margin of dicotyledonous or broad-leaved plants and of the tips of monocotyledonous or grass-like, parallel veined plants. The affected areas may be light tanned to brown or dark red in color. The death of tissues occurs when the accumulation of fluoride reaches the toxic level for the species in question. Since the leaf accumulates fluoride, chemical analysis is used as an aid in determining sources of pollutants and the extent of an affected area around the source. The whole leaf surface absorbs fluoride which is then gradually translocated to and accumulated in the leaf margins or tips. Exposures required may be in terms of weeks or months and concentrations usually occur at 1 ppb or less. The difference of concentration of fluoride in affected leaves between the center of the leaf and its margin may vary by a factor up to 100. Concentrations in leaf tissues is expressed in terms of parts per million of dry weight of the leaf. The accumulation of as little as about 30 ppm fluoride in gladiolus is enough to cause injury, whereas accumulation of 5,000 ppm in cotton does not cause any injury. The normal fluoride content of many plants is between 2 and 20 ppm;

some exceptions noted are tea, which may have a natural content
from 50 to 90 ppm, and camellia having up to almost 2,000 ppm.
Thus when the normal fluoride contents are known, elevated levels
indicate that air pollution may be a problem. One must exercise
considerable judgment on symptoms alone because excess salt accumulation, water stress, and some mineral deficiencies can cause
similar markings.

Needles of some pine trees are quite susceptible and large
areas near Spokane, Washington, have been devastated in the past
by hydrogen fluoride. Affected pine needles are shed in one to
two years instead of the normal three to five years. Injury
develops when needles accumulate according to the following figures:
young needles that are partly expanded--15 to 20 ppm, freshly
expanded needles--30 to 35 ppm, three-month expanded needles--80
ppm, three-year old needles--up to 460 ppm.

The question of growth suppression fluoride is not clear.
Some workers investigating gladiolus and tomatoes indicate that
there is no effect on growth in the absence of injury to the leaves.
Other workers who have exposed citrus have indicated that there is
a definite growth suppressing effect without any injury to the
leaves. This question becomes very important in fluoride pollution
control because of the fact that plants accumulate fluoride from
such very small concentrations in the air.

Even without death to leaf tissues or growth suppression,
accumulation of fluoride in tissues of forage crops becomes important to the animals which feed on those plants. Thus, alfalfa
may show no outward signs of being affected by fluoride, but if
fluoride is accumulating at low levels in those leaves, the accumulated fluoride is transferred to the animal's body where it further
accumulates to cause the disease known as fluorosis. Teeth become
mottled, gums recede, teeth become loose, joints are stiff, the
animal loses its appetite and there is no gain in weight; sometimes there is even death of the animal. Thus because of its
affect on both plants and animals, fluoride is often considered
one of the most important pollutants in today's economy.

Acid aerosol--In foggy or misty weather sulfur dioxide emissions
may form acid droplets which settle on leaves. As these droplets
dehydrate with time the acid becomes sufficiently concentrated to
burn leaf tissues and cause small discrete spots, usually confined
to the upper surface of the leaves.

Dust--Very little research has been done on dust effects and
most of this work has been done in Germany, although a new project
has been underway in California for a little over four years. The
mere presence of dust may affect the market value of ornamental

plants or reduce the market grade of forage crops. Certain cement-
kiln dusts have been shown to be toxic to plant leaves when deposited
in the presence of free moisture. The alkyline solution resulting
from such deposits is able to penetrate the leaf cuticle and burn
the tissues underneath. The long range effect on plants is that
growth is significantly reduced. Needles of affected pine trees
may fall prematurely. Dry dusts apparently do not have this affect
but may interfere with the normal photosynthesis by the fact that
they reduce the light reaching the leaf. The long range effect of
this has yet to be determined.

Some field observations indicate an indirect effect of dust on
plants. In one instance alfalfa downwind from a cement plant was
covered with dust. The plants were badly stunted and were also
heavily infested with aphids. A short distance away out of the
path of dustfall the plants were of normal height and free of aphids.
It was postulated that the dust eliminated the natural parasites
of the aphid thereby allowing them to increase to the point that
they were able to do serious injury and cause stunting of the
alfalfa plant.

Ecological Considerations

In addition to the specific effects on crop and forest plants
noted in this discussion there is now increasing concern about
the long term pollution effects on the natural vegetation over
the earth. In other words, are certain species over wide areas
being eliminated and others encouraged to our detriment or benefit?
This subject is discussed in some detail by Professor Treshow in
the symposium listed in the following bibliography.

Selected Bibliography

1. Adams, D. F. The effects of air pollution on plant life.
        Am. Med. Assoc. Arch. Ind. Health, 14: 229-45 (1956).
2. Adams, D. F. Recognition of the effects of fluorides on
        vegetation. J. Air Pollution Control Assoc., 13: 360-
        62 (1963).
3. Berry, C. R., and Ripperton, L. A. Ozone, a possible cause
        of white pine emergence tipburn. Phytopathology 53: 552-
        57 (1963).
4. Brewer, R. F., Sutherland, F. H., and Guillemet, F. B. Effects
        of various fluoride sources on citrus growth and fruit pro-
        duction. Environ. Sci. and Tech. 3: 371-381 (1969).
5. Darley, E. F. Studies on damage to vegetation from cement-
        kiln dust. J. Air Pollution Control Assoc. 16: 145-50
        (1966).
6. Darley, E. F., Dugger, W. M., Mudd, J. B., Ordin, L., Taylor,
        O.C., and Stephens, E. R. Plant damage derived from auto-
        mobiles. Am. Med. Assoc. Arch. Environ. Health, 6: 761-
        770 (1963).

7. Darley, E. F., and Middleton, J. T. Problems of air pollution in plant pathology. Ann. Rev. Phytopath., 4: 103-118 (1966).

8. Darley, E. F., Nichols, C. W., and Middleton, J. T. Identification of air pollution damage to agricultural crops. Calif. Dept. Agric. Bul. 55: 11-19 (1966).

9. Davidson, O. W. Effects of ethylene on orchid flowers. Proc. Am. Soc. Hort. Sci., 53: 440-66 (1949).

10. Haagen-Smit, A. J., Darley, E. F., Zaitlin, M., Hull, H., and Noble, W. Investigation on injury to plants from air pollution in the Los Angeles area. Plant Physiol., 27: 18-34 (1952).

11. Hepting, G. H. Air pollution impacts to some important species of pine. J. Air Pollution Control Assoc., 16: 63-65 (1966).

12. Middleton, J. T. Photochemical air pollution damage to plants. Ann. Rev. Plant Physiol., 12: 431-48 (1961).

13. Miller, Paul R. Photochemical oxidant injury and bark beetle infestation of ponderosa pine: III. Effect of injury upon oleoresin composition, phloem carbohydrates, and phloem pH. Hilgardia, 39(6): (1968).

14. Mudd, J. B. Responses of enzyme systems to air pollutants. Am. Med. Assoc. Arch. Environ. Health, 10: 201-06 (1965).

15. Ordin, L. Effect of air pollutants on cell wall metabolism. Am. Med. Assoc. Arch. Environ. Health, 12: 189-94 (1965).

16. Rich, Saul. Ozone damage to plants. Ann. Rev. Phytopath., 2: 253-66 (1964).

17. Special Jubilee Symposium, 1958. Air pollution with relation to agronomic crops. Seven papers by several authors discussing effects of air pollutants on vegetation. Agron. J., 50: 244-568 (1958).

18. Symposium: Trends in air pollution damage to plants. Phytopathology 58: 1075-1139 (1968). These pages cover 7 articles by various authors on the effects of air pollution on plants.

19. Stephens, E. R., Darley, E. F., Taylor, O. C., and Scott, W. E. Photochemical reaction products in air pollution. Intern. J. Air Water Pollution 4: 79-100 (1961).

20. Taylor, O. C., and Eaton, F. M. Suppression of plant growth by nitrogen dioxide. Plant Physiol., 41: 132-35 (1966).

21. Taylor, O. C. Importance of peroxyacetyl nitrate (PAN) as a phytotoxic air pollutant. J. Air Poll. Control Assoc. 19: 347-351 (1969).

22. Thomas, M. D. Gas damage to plants. Ann. Rev. Plant Physiol., 2: 293-322 (1951).

23. Thomas, M. D. Effects of air pollution on plants. In Air Pollution, World Health Organ.: Monograph Ser., No. 42: 233-78 (1961).

24. Thompson, C. Ray. Effects of air pollution on lemons and navel oranges. Calif. Agric. 22: 2,3 (1968).

WASTE DISPOSAL

Victor P. Osterli

Agricultural Extension Service

University of California at Davis

The purpose of this paper is to outline the problems of solid waste disposal and to discuss some of the alternatives for their solution. Emphasis will be given to agricultural wastes since some of the problems associated with their alleviation appear to be less well understood.

I. Pollution Problems

A recent study by the California Department of Public Health summarizes the environmental problems identifiable with solid wastes in the state as follows:
"(1) The creation of severe disease-carrying domestic fly and rodent densities as a result of poorly managed solid wastes;
(2) Air pollution and smoke nuisance problems from widespread burning of solid wastes;
(3) Pollution of ground and surface waters from inadequate solid waste disposal systems; and,
(4) The proliferation of public nuisances from odors, smoke, fire hazards, and unsightliness."

In addition there are other more indirect effects of present day solid waste disposal practices which include the rising economic burden, degradation of property values, and the erosion of esthetic values.

Pollutants are either the unwanted by-products of our activities or spent substances which have served intended purposes. Man has always produced pollutants and until more recently, natural

processes were sufficient to change most of them into harmless or
beneficial substances. Today, however, they are produced in such
large quantities that the capacity of the air, the land, and the
water to absorb them is often exceeded. This is the age of the
"no deposit, no return."

It is not possible to separate solid waste disposal from
the air and water pollution problems. Air pollution may be aggravated by the burning of solid waste refuse; the household grinding
of garbage eliminates a solid waste but increases the liquid waste
load; the disposal of certain solid wastes by landfill may produce
a water pollution problem. On the other hand, the abatement of an
air pollution problem may produce an additional solid waste problem.

During 1967 a total of 71.5 million tons of solid waste was
produced in California. This is broken down as follows:

"Municipal Wastes	Tons	
Residential	8,866,000	
Commercial	9,717,000	
Demolition	2,988,000	
Special	1,343,000	
		22,914,000 (22.9)
Industrial Wastes		
Food Processing	2,127,000	
Lumber	7,993,000	
Chemical & Petroleum	464,000	
Manufacturing	3,103,000	
		13,687,000 (13.7)
Agricultural Wastes		
Manure	21,809,000	
Fruit & Nut Crops	2,361,000	
Field & Row Crops	10,731,000	
		34,901,000 (34.9)
1967 Total		71,502,000 (71.5)"

Based on a population of 19.5 million people, the total per
capita waste production in California exceeds 20 pounds per day,
broken down as follows:

Municipal wastes   ---   6.5 pounds
Industrial wastes   ---   3.9 pounds
Agricultural wastes --   9.8 pounds

A frequently quoted national estimate of the unit output of
municiapl solid wastes is 4.5 pounds per capita per day which by

1980 is expected to be 5.5 pounds per day. California, at 6.5
pounds per capita per day, is already well over the projected 1980
figure.

Composition of municipal solid waste is highly variable.
One source has indicated the following general composition:

Type	Percent
Paper	45
Grass, brush, cuttings	15
Garbage	12
Ashes	10
Metallics	8
Glass and ceramics	6
Miscellaneous	4

Solid waste disposal methods currently in use in California
generally involve some burning or burying and, in a few instances,
organic conversion. Municipal wastes are in nearly all cases disposed of by burning or burying in landfills. Industrial wastes
are commonly disposed of in small landfills on plant property,
transported to other landfills, or burned. Agricultural wastes
are disposed of mainly by spreading on land or burning.

II. Solutions and Alternatives

Refuse disposal management involves planning, organizing,
directing and controlling the various phases of the process. Basic
management principles are the same for cities of all sizes. It is
only their application that varies from place to place.

Because solid wastes include not only the output of households
and municipalities but also includes the discards of business, industry and agriculture, there is variety among the handling procedures. When material reaches the discard point, its handling can
be divided into three steps:
    -- Collection, which includes storage and transfer, as
well as pick up;
    -- Processing, which may take a variety of forms including
the salvage of usable and useful portions; and
    -- Disposal, which includes any treatment necessary to make
the disposition effective.

The handling of household wastes involves all three of these
steps in a coordinated way. Many agricultural wastes, on the
other hand, are still disposed of at or near their points of origin.

Methods of ultimate disposal are limited. These include:
    Incineration of combustibles

Composting
Animal feeding (recycling)
Disposal at sea of non-floating materials
Landfill

It is beyond the limits of this paper to review each in detail. There are in California 716 general-use disposal sites in
operation receiving about 19.5 million tons of refuse per year.
Just over 500 (71%) of these are burning dumps, while 209 (29%) are
controlled landfills. Only 67 disposal sites are rated as acceptable
sanitary landfills. While sanitary landfilling is not without
problems, the use of this method is likely to increase markedly.

A recent report on solid waste management in San Mateo
County summarizes today's critical management needs as follows:
"The study indicates the need for policy guidance to the various
interests involved in resolving the disposal problems... It is
apparent that the many separate waste management programs and smaller systems of the past are obsolete. Although the critical problems facing us now and in the future are a local responsibility,
they cannot be effectively resolved on a local basis except through
a fully coordinated, local-regional disposal system."

Because incineration of wastes merely involves a change of
state --- primarily of solids to gases and tiny particulates ---
the land and the ocean in reality represent the only permanent sinks
into which wastes ultimately can be disposed. Flowing water and
even the atmosphere have only limited reservoir capacity. This is
not to say that incineration should not be utilized. Its continued
use can be justified only under conditions where the combustible
refuse is burned without nuisance.

Combustion-generated air pollution results from open burning
for the disposal of forest and crop residues. These result from
harvesting and cultural operations and from accumulations resulting
from pest damage or natural deterioration. Forest and crop residues
serve as a reservoir of plant diseases, insects and other pests.
Some of these residues, if properly managed, can be left in the area
produced. Others, such as logging debris, are serious problems
because of the fire hazard they create or because they are hosts
for diseases and insect pests.

Agriculture recognizes that some of its operations do contribute to air pollution. The aggravation and the offensiveness of
any pollutants caused by long-used agricultural operations result
more from the increasing occupancy of former open land by people
than from any increase in these operations themselves. Previously
accepted practices of agricultural husbandry have now become objectionable sociological problems.

Technological advances have demonstrated that, in some instances, alternative methods to the use of fire for agricultural waste disposal are advantageous. Many of these techniques have been adopted and have become the accepted cultural practice when economical and practical.

The most common method in disposing of field crop, row crop, and fruit and nut crop residues is to incorporate the material directly into the soil by discing, or by chipping or shredding for subsequent working into the soil. In the case of non-tillage culture, the material is left to decay on the soil surface.

Because agriculture is being examined with increasing vigilance as a source of pollution, the University of California in copperation with the agricultural community has undertaken studies to determine the extent to which agriculture may be a contributor.

Burning surveys first made in 1959 and 1960 have been updated for some of the counties. During a 7-year period, 1959 to 1965, there was a 9 percent reduction in the volume of deciduous fruit, grape, and nut tree pruning and dead trees disposed of by burning in the counties comprising the Bay Area Air Pollution Control District. Some of this was due to a change in land use from orchards to urbanization or industrial development. In addition commercial flower growers have discontinued open burning; most grape prunings are incorporated into the soil; chemical and mechanical control of rangeland brush is displacing control burns. Orchard prunings and attrition losses constitute the largest group of wastes still burned in this area. Approximately 25 percent of these are disposed of by means other than burning. The survey further showed that most of the agricultural burning (73%) was done during the period of the year when oxidant levels were normally low, namely the winter and early spring months.

In Kern and Tulare Counties 80 percent and 74 percent, respectively, of the agricultural wastes are disposed of by means other than burning. If the additional 330,000 acres of cotton, dry and green beans, melons, sugar beets, potatoes, tomatoes and several other vegetable crops whose residues are normally disposed of through incorporation into the soil are included, only about 10 percent of the crop residues in Kern County and approximately 21 percent in Tulare County are still burned.

A recently completed survey of agricultural burning in the Sacramento Regional Area Planning Commission* jurisdiction shows

---

*The Sacramento Regional Area Planning Commission includes all of

that only 15 percent of the agricultural crop residues are burned.
The remaining 85 percent is mainly disposed of by incorporation
into the soil. There is an additional 200,000 acres of grain sor-
ghums, safflower, seed crops, sugar beets, tomates and other vege-
tables whose residues are disposed of by incorporation into the
soil. In the total area of the five counties of El Dorado, Placer,
Sacramento, Sutter, and Yolo, approximately 80 percent of all of
the residues of field and row crops and fruit and nut crops are
disposed of other than by burning.

Attempts to Clarify the Problem

Public concern for the growing level of air pollution in the
Sacramento and San Joaquin Valleys and a lack of specific data on
the contribution by agricultural burning to air pollution, led the
Department of Public Health and the University of California Division of Agricultural Sciences to investigate the problem. Field
investigations were carried out during 1959 and 1960 with the cooperation of federal, state and local agencies. During 1959, several
air sampling programs were undertaken in conjunction with the burning of field crop residues and slash from range improvement projects.
These studies provided valuable qualitative information on the
chemical nature of emission from individual fires. They showed
that late fall agricultural burning in the lower Sacramento Valley
contributed to particulate pollution in that area, but there was
no indication that photochemical pollution in Sacramento was increased due to burning.

Further Studies Initiated

The foregoing field studies did not permit an evaluation of
emissions in terms of pollutant per unit of waste burned. Experiments were subsequently conducted where known weights of various
agricultural wastes were burned in a specially designed tower that
simulated open burning. Adequate sampling of the effluents was
possible.

The effluent from 123 fires of various agricultural wastes
was studied. These included deciduous fruits, grapes, nuts, field
crops, straw and stubble, native brush from rangeland and such
miscellaneous materials as chips of douglas fir, redwood, and cotton
debris.

The burning tower with its associated apparatus and analytical

---

Sacramento and Yolo Counties, all of El Dorado and Placer Counties
(except the Tahoe Basin) and the portion of Sutter County south of
the Bear River.

instruments made it possible to determine the pounds of effluent per ton of waste material burned. The results obtained show that the maximum expected emissions of hydrocarbons (saturates--except methane, olefins, and acetylenes) per ton of waste when burning fruit tree prunings, rice straw, barley straw, and dry native range brush would be approximately 14, 9, 18, and 7 pounds, respectively. Green native brush produces as much as 36 pounds of hydrocarbon per ton, but this type of waste constitutes a relatively small portion of the total material burned in a range improvement program. In comparison, automobile exhaust when uncontrolled produces about 130 pounds of the same hydrocarbons per ton of fuel. It is evident from these studies that the contribution of total hydrocarbons from agricultural burning is very small when compared to the automobile exhaust.

Particulate Emissions from Agricultural Burning

Quantitative evaluation of particulate emissions from the burning of agricultural wastes only recently has been developed. A cooperative study by Oregon State University and University of California researchers sponsored in part by a research grant from the National Air Pollution Control Administration showed that particulate emissions from burning grass straw and stubble averaged 15.6 pounds per ton of fuel burned. The particulate matter collected in the field studies averaged 15.5 pounds of particulate per ton of fuel burned.

Soil Incorporation Effects Being Studied

An extensive study supported by a research grant from the United States Public Health Service Solid Wastes Program is underway. This project deals with the effects of incorporation of orchard and vineyard prunings and straw and stubble into the soil upon the rate of decay of the shredded plant material. Other incorporation effects being studied include: plant disease organisms and other soil microflora, certain insect pests, and various soil properties. Researchers from the Riverside, Berkeley and Davis Campuses are involved. Results of this project should tell us whether these agricultural wastes can be incorporated into the soil without adverse effect upon future production.

Controlled Burning Helps Entire Community

Since 1945, California cattlemen and private rangeland operators have control-burned 2,306,551 acres under permit. This involves the use of fire for the removal of "less desirable" woody vegetation followed by reseeding with forage species to provide more feed of better quality for domestic cattle. Such burning also has increased both the amount and quality of feed and has improved

the habitat for wildlife. Conducted by private landowners under
supervision of the California Division of Forestry*, this program
also provides additional benefits to adjacent urban communities in
the form of increased water yield, reduced fire hazard, and improved
access for sportsmen as well as cattlemen.

Less Use of Fire for Range Improvement

Chemical and mechanical control of woody species is significantly reducing the acreage that is control-burned each year.
During the decade, 1948 through 1957, a total of 1,309,984 acres
was control-burned. During the subsequent 10-year period, 1958-
1967, this was reduced to 859,174 acres. Thus the acreage burned
was reduced from an average of 131,000 acres to 86,000 per year.
During the latter period, 1958-1967, private landowners cleared
100,000 acreas by discing, bulldozing, or other mechanical means
and also cleared by chemical treatment an additional 117,600 acres.
The latter treatment utilizes foliar or basal sprays, cut-surface
treatment, or soil sterilants. During this same period 88,900 acres
of public lands were similarly treated.

From the standpoint of wildfire prevention it is difficult
to assess the value of open areas created within brush fields by
the range improvement program. Analysis of California Division of
Forestry wildfire figures indicates the wildfire prevention aspect
is significant. State Board of Forestry policy recognizes controlled
fire as a method appropriate to the development, management, and
conservation of natural resources of California. Hopefully, the
controlled burning program administered by the California Division
of Forestry may be continued as a permissive tool in wildland
management. Controlled fire, with herbicides and mechanical means
of brush removal, is an important land management tool for the private landowner as well as for the public land manager. Control
burns are generally at an elevation above troublesome inversion
layers--very limited use is made of it in southern California.

Many Burning Rules, Regulations Already in Effect

Agriculture in some county air pollution control districts
operates under open burning restrictions. These permit burning
only when the inversion base and the maximum mixing height are at
prescribed levels and only under specific wind velocities. Counties
without air pollution control districts have regulated emissions
from orchard heating operations by county ordinance.

---

*Administration of range improvement burning is provided by Sections
4490-4494 of the Public Resources Code.

Following several years of discussion, the Bay Area Air Pollution Control District and growers reached agreement on rules and
regulations regarding burning of agricultural wastes. The burning
of prunings and attrition losses from tree fruits, nuts and grapes
are permitted during the period from December 1 through April 30
under the following conditions:

> "(a) The height of any temperature inversion, as
> measured at the Oakland Airport by the United States
> Weather Bureau, at 1200 Greenwich Mean Time, is not
> less than 2500 ft. m.s.l.; or the surface temperature,
> measured at the site of burning, will, by standard
> methods of calculation, have caused the inversion to
> reach 2500 ft. m.s.l. or greater. (b) No burning shall
> be done before 9:30 a.m. P.S.T. (c) No additional
> pruning material or ignition fuel shall be ignited
> nor shall any pruning material or ignition fuel be
> added to any fire after two hours before sunset.
> (d) No pruning material or ignition fuel shall be
> be ignited nor shall any pruning material or ignition
> fuel be added to any fire when the wind velocity is
> less than 5 miles per hour. (e) The moisture content
> of material burned shall not exceed 30 percent wet
> basis. (f) Wind direction at the site shall be
> such that the direction of smoke will be away from
> populated areas to minimize local nuisance caused
> by smoke and particulate fallout. (g) No burning
> shall take place without the permission of the Air
> Pollution Control Officer."

During the first season these regulations were in force
(1967-68), it was necessary to telephone the district office to
obtain permission to burn. Permission was granted, usually subject to a minimum permissible temperature at the burn site.

Administrative procedures were simplified during the 1968-
69 season while basic controls remain unchanged. Farmers listened
for the district's announcement stating whether it was a "burn day"
or a "no-burn day." The announcement was broadcast each day during
the burn season on local radio stations. If the day was declared
a "burn day" farmers burned. It eliminated having to call the
district office to register and receive permission. There were
124 burn days out of the 152-day burn season the first year while
during the 1968-69 season there were 104 burn days and 47 no-burn
days.

Rice Growers Try Volunteer Program

Rice growers in the Sacramento Valley, industry representa-

tives, California Farm Bureau Federation officials, and University
of California researchers and Agricultural Extension staff members
are studying all aspects of the air pollution problem. One attempt
to alleviate the situation is to burn stubble and straw only under
meteorological conditions favorable to the best dispersal of smoke.

The Sacramento office of the U. S. Weather Bureau augmented
their weather forecasting service in the fall of 1968. The service
began on October 1 and continued into November. It was designed as
an experimental program to assist rice growers and others of the
agricultural community in planning their agricultural burning operations during the fall season. The service was made available for
a short period during the spring also.

The new service, termed "Ventilation Forecasts for Agriculture" was aired on local radio stations. The forecasts included
wind information and a one-word term which indicated dispersal
characteristics in the lower few thousand feet of the atmosphere
during the afternoon hours. The terms "excellent," "good," "fair,"
or "poor" were used to indicate how effectively any smoke would be
dispersed throughout the air. If the ventilation forecast indicated
"excellent," it meant "best dispersal of smoke." "Poor" conditions
meant smoke would tend to concentrate nearer the ground. When
ventilation conditions differed within the Sacramento Valley, separate forecasts were provided using well know geographical boundaries
as area delineators.

The regular daily weather forecasts and 5-day outlooks is-
sued by the Weather Bureau through the news media also were used
by growers in planning their various agricultural operations. Use
of the forecasts was strictly voluntary. It was an opportunity to
test the possible effectiveness of the service and to demonstrate
to the entire community agriculture's desire to assist in reducing
the problem.

The Oregon Environmental Quality Commission pursuant to 1969
legislation has adopted rules and regulations for field burning for
a 9-county area in the Willamette Valley. This is applicable during
the summer agricultural burning season, July through October. Field
burning is permitted when meteorological conditions are satisfactory
for smoke dispersal. Certain types of atmospheric conditions have
been classified "marginal" and the specified type and extent of
burning allowed on each type of "marginal" day has been established
(see table).

The Willamette Valley of Oregon is an important grass seed
producing area. Research has shown that burning crop residues in
fields of cool season perennial grasses after harvest helps the
next year's seed crop. Although burning was originally intended

SCHEDULE OF MARGINAL DAYS AND CORRESPONDING
BURNING RESTRICTIONS

Class		Meteorological Conditions	Allowed Burning*	Burning Hours** Begin	End
UNRE-STRICTED	1	Forecast Maximum Mixing Depth 5500 feet or greater	No restrictions on type of burning	Time Mixing depth is forecast to reach 3000'	Sunset
MARGINAL	2	Forecast Maximum Mixing Depth 5000-5400 feet	Annual and Perennial grass seed fields used for grass seed production; cereal grain fields	Time Mixing depth is forecast to reach 3000'	Sunset
	3	Forecast Maximum Mixing Depth 4500-4900 feet	Annual and perennial grass seed fields used for grass seed production	"	"
	4	Forecast Maximum Mixing Depth 4000-4400 feet	Perennial grass seed fields used for grass seed production	"	"
	5	Forecast Maximum Mixing Depth 3000-3900 feet	Perennial grass seed fields used for grass seed production	1 P.M.	6 P.M.
PROHIB-ITED	6	Forecast Maximum Mixing Depth less than 3000'	Burning prohibited except propane flaming, where combustion is nearly complete		

(table continues)

*Allowed Burning - Note that "other burning," which includes pastures, fence rows and ditch banks (including those around grass and grain fields), agricultural land clearing debris, brush, etc., is not allowed under marginal conditions.

**Burning is to be initiated and completed between these hours.

SCHEDULE OF MARGINAL DAYS AND CORRESPONDING
BURNING RESTRICTIONS (Continued)

Class	Meteorological Conditions	Allowed Burning*	Burning Hours** Begin	End

FURTHER RESTRICTIONS

Whenever visibility at Salem or Eugene airport, as observed by the U. S. Weather Bureau in the NW quadrant is reduced to 6 miles or less by smoke or haze for two consecutive hours, or to 3 miles or less at any time under prevailing relative humidities of 70% or less on any day, the following day shall be prohibited or classed marginal and days shall be classed as follows:

Class 3   Mixing depth - 5500' or more
Class 4   Mixing depth - 4500' or more
Class 5   Mixing depth - 4000' or more
Prohibited   Mixing depth - less than 4000'

The staff of the Sanitary Authority or its successor agency may authorize burning in excess of that permitted by the schedule where conditions in their judgment warrant it, or, by express written permit, burning on an experimental basis, and may also, on a fire district by fire district basis, issue limitations more restrictive than those contained in the schedule, when in their judgment it is necessary to attain air quality.

---

as a control measure for blind-seed disease in perennial ryegrass, the benefits of open field burning for other grass seed crops were noted.

All grass species do not react the same to open field burning. Yields of fine fescues and bentgrasses drop sharply if burning is not done annually. In other grasses, the effects of yearly burning on seed yield are not quite so dramatic. Burning is not just a way to dispose of residues. It is an important management tool.

The timber industry, like any other, is faced with the necessity of increasing the percentage of raw materials converted to marketable products in order to maintain a competitive position. Paradoxically, increasing use of residues has made residue disposal more of a community nuisance. A properly operated teepee burner of the right size is relatively smoke-and cinder-free when burning the residues from a mill without by-product recovery. However, even the best burner operating techniques are not adequate to prevent smoke and cinders when the usable slabs, edgings, and other large chunks have been removed. The sawdust, shavings and bark remaining tend to settle in a compact mass on the burner floor--making it virtually impossible to attain the temperature necessary for a nuisance-free operation.

Current Status

With the progress at putting into practice alternative
methods for crop residue management and agricultural waste disposal, what is the situation today? There is more to crop residue
management than merely incorporating material back into the soil.
At least three factors are especially significant. These are the
problems of pests (either diseases or insects), economics and availability of an alternative method.

Herbicides have to a large extent replaced open burning
along ditch-banks, roadsides, and around farm headquarters. The
floriculture industry has discontinued the open burning of green-
house wastes. These are disposed of by hauling to cut-and-fill
dumps. The disposal of vineyard prunings by burning is no longer
a common practice. The only grape prunings still burned are in
older vineyards where close planting (8x8 feet or less) makes it
difficult to use other methods.

The incorporation of prunings into the soil has become an
accepted practice in southern California and in the southern San
Joaquin Valley. The most common methods are either to incorporate
the material directly into the soil by discing or to chip or shred
for subsequent working into the soil. In the case of non-tillage
culture, the material is shredded or chipped and left to decay on
the soil surface. These practices are not utilized to the same
degree in fruit producing areas of the Sacramento and Santa Clara
Valleys. Any wood too large for the chippers or shredders has to
be separated out, handled, and disposed of in some other way.

Burning of straw and stubble of cereal grains on upland ter-
race soils under summer fallow culture is no longer a common practice. These areas are often grazed by cattle or sheep reducing
the volume of material necessary to turn into the soil.

Although there is a reduction in burning upland grains,
that reduction is being partially offset by double cropping on
intensively farmed, irrigated land. The economic necessity for
maximizing the annual net income per acre is responsible. Involved
are such crops as barley, oats, or wheat, which are harvested in
early summer. They are followed by a second crop planted just as
soon as the stubble and straw wastes are removed, the ground is
irrigated and otherwise properly prepared. Crops requiring precision planting and several cultivations present a special problem.
Incorporating the straw and stubble into the soil before planting
a crop such as beans or milo presents a mechanical problem in preparing the soil, in the subsequent cultivating operations, and with

certain insect pests. Complete removal of all straw and stubble
is necessary for preparation of the beds and cultivation for such
a crop as lettuce.

        Utilization of crop residues by livestock is still a common
practice. Almond hulls are a highly nutritious livestock feed and
almond shells make excellent bedding for cattle. Recent experiments
indicate that adding protein to rice straw and steaming it makes it
97 percent as good a feed as alfalfa. There are no figures, however,
regarding the economics of such a procedure.

Burning Only Solution in Some Cases

        The University of California conducts research into the
biology as well as the control of agricultural pests. Its studies
include both chemical and biological control measures. Ways and
means are sought to integrate the best of these two programs for
more efficient and safe control. Only rarely do the scientists
resort to burning as a means of abatement of agricultural pests.
However, in a limited number of cases burning as a control measure
is necessary.

        A case in point is the stem rot disease of rice (Sclerotia
oryzae). It has been found in many of the California rice producing
counties and its rate of spread and build-up is not yet fully known.
Yields have been reduced by over 50 percent. Control of this disease in the southern United States rice producing areas is accomplished by burning infected fields. The fruiting bodies (sclerotia)
live on decaying organic matter.

        Disposal of animal wastes is a major and ever-growing pro-
duction problem for the agriculturist. A comparison of the wastes
voided by man and by the animals he raises illustrates the magnitude
of the problem. A cow generates as much manure as 16.4 humans, 1
hog produces as much waste as 1.9 people, and 7 chickens provide a
disposal problem equivalent to that created by 1 person.

        Animal manures also constitute a major portion of the agri-
cultural waste problem in California. Compounding the problem are
the tremendous changes that have taken place in livestock handling.
Some farmers still produce a significant percentage of their livestock feed. Today some 2 million head of cattle go through California's nearly 500 commerical feedlots each year. The average size
is 4000 head, but ranges from 500 to 50,000 head. Many modern dair-
ies are strictly drylot operations and have 200 or 300 cows on 10
or so acres. The farm poultry flock has been replaced by the operation which confines 100,000 chickens in cages, 1/2 square foot of
floor space per bird.

        The benefits from the use of animal manures have been

ascribed both to the plant nutrient contained and to the organic matter supplied to the crop land. Their nutrient value comes primarily from the nitrogen, phosphorus, and potassium contained in the materials. Poultry manures have the highest in NPK content while dairy and feedlot manures tend to be very much lower. The nitrogen content may decrease rapidly with volatilization of ammonia or leaching of potassium and ammonium compounds with rainfall. Manures vary greatly in composition depending upon the length and kind of storage and on the porportionate admixture of foreign materials such as bedding and soil.

Animal manures have value only when applied to a soil where a responsive crop is present and where a need for the nutrient supplied by the manure exists. Value can only be ascribed where the nutrients are actually needed. Economic comparisons with commercial fertilizers should be made on an applied basis at point of use. Evaluations should be made on the basis of crop performance comparing manures with standard fertilizer materials readily available and commonly used by commercial agriculture.

Care must be taken in use of manure since heavy applications can result in nutritional imbalances, such as induced magnesium or zinc deficiencies. Excess nitrogen applied on crops to be grazed may raise the nitrate content to toxic levels for animal use. Though processing of animal manures for retailing will continue, it is not likely soon, if ever, to constitute any significant channel of disposal.

Agricultural waste management is an exceedingly complex problem. There is no easy solution. There is no single answer. Progress has been made with livestock waste disposal including biological treatment, materials handling, and studying the economics of using manures as plant nutrients. Progress has been made on disposal of crop residues and wastes by methods other than burning. Technological advances have demonstrated that alternative methods to the use of fire for agricultural waste disposal are advantageous in some instances. These have become the accepted cultural practice, when economical.

In addition to the research effort, the educational program has also concerned itself with acquainting the agricultural community with the need to develop and initiate alternatives to open burning.

The agricultural community has demonstrated its willingness to adopt alternative methods and work toward the development of realistic rules and regulations for the use of open burning as a management tool. Whatever the motivating force for regulation, control or suggested alternatives to burning, the farmer is concerned that any such action be based upon complete evaluation of all the facts. The search for alternatives is a continuing effort.

Selected References

1. County Burning Surveys. University of California, Agricultural
   Extension Service, Berkeley. 1960.
2. A Field Study of Air Pollution in the Lower Sacramento Valley
   During the 1960 Fall Season. California Department of
   Public Health.
3. Proceedings, National Conference on Solid Waste Research.
   American Public Works Association. December 1963.
4. Restoring the Quality of Our Environment. Report of the Environmental Pollution Panel, President's Science Advisory
   Committee. November 1965.
5. Solid Wastes Management. Proceedings of the National Conference,
   University of California, Davis. April 1966.
6. Management of Farm Animal Wastes. Proceedings, National Symposium on Animal Waste Management. Michigan State University. May 1966.
7. A Progress Report Relating to the Disposal of Agricultural
   Solid Wastes in the Bay Area. University of California
   Agricultural Extension Service and Air Pollution Research
   Center. July 1966.
8. Contribution of Burning of Agricultural Wastes to Photochemical
   Air Pollution. Air Pollution Control Association Journal,
   Vol. 16, No. 12. December 1966.
9. Municipal Refuse Disposal. American Public Works Association.
   1966.
10. Proceedings, California Tall Timbers Fire Ecology Conference.
    November 1967.
11. Agriculture and the Quality of Our Environment. American
    Association for the Advancement of Science. 1967.
12. Agricultural Utilization of Sewage Effluent and Sludge - An
    Annotated Bibliography. Federal Water Pollution Control
    Administration. January 1968.
13. Waste Management Research and Environmental Quality Management
    Hearings before the Subcommittee on Air and Water Pollution
    of the Committee on Public Works - United States Senate.
    May, June and July 1968.
14. Air Pollution and the Community. University of California
    Agricultural Extension Service, Tulare County, Visalia,
    California. July 1968.
15. Air Pollution and the Community. University of California
    Agricultural Extension Service, Kern County, Bakersfield,
    California. September 1968.
16. Status of Solid Waste Management in California. California
    Department of Public Health. September 1968.
17. Air Pollution and Agriculture Today. California Agriculture,
    University of California, Berkeley, California. September
    1968.

18. Ventilation Forecasts for Agriculture. Farmers' Rice Coopera-
    tive Coop Newsletter, Vol. 4, No. 2. October 1968.
19. Agricultural Waste Management in the Future. Agricultural En-
    gineering. December 1968.
20. Control of Agriculture-Related Pollution. A Report to the Presi-
    dent, submitted by the Secretary of Agriculture and the Di-
    rector of the Office of Science and Technology. January,
    1969.
21. Wood Processing Residue Disposal. California Agriculture.
    University of California, Berkeley, California. March 1969.
22. Air Pollution and Crop Residue Management - Greater Sacramento
    Area. University of California Agricultural Extension Ser-
    vice. May 1969.
23. Recommended Changes in Disposal of Solid Wastes, City of Davis.
    June 1969.
24. Burning Fields Boosts Grass Seed Yields. Crops and Soils.
    Vol. 21, No. 8. June-July 1969.
25. Emissions from Burning of Grass Stubble and Straw. Air Pollu-
    tion Control Association Journal. Vol. 19, No. 7. July
    1969.

CENTRAL POWER PLANTS

Robert F. Sawyer

Department of Mechanical Engineering

University of California, Berkeley

INTRODUCTION

Central power plants have the dubious distinction of being third in their contribution to air pollution in the United States, trailing motor vehicles and industry. Power plants emit approximately 25% of the particulates, 46% of the oxides of sulfur, and 25% of the oxides of nitrogen which are released each year into the atmosphere in the United States. They are minor contributors of carbon monoxide and hydrocarbons.

Since power plants are few in number (in comparison with automobiles) they were identified early as concentrated sources of air pollutants. Initial concern was with particulates and sulfur dioxide. More recently some notice has been given to their contribution of nitric oxide.

SOURCES OF ELECTRICAL POWER

The electric utilities are a major consumer (and producer) of energy in the United States. The growth of the electric utilities has exceeded and is projected to exceed the growth of other major energy consuming sectors. The total and percentage energy consumption for the several sectors are summarized in Table 1.

The electric utilities may be divided according to the energy source employed for the generation of electricity. Although a dramatic increase in the use of nuclear power is forecast, the absolute growth in generating capacity of fossil fuel plants will exceed that for nuclear power. Recent and projected distributions of electric energy sources are presented in Table 2. Longer

range predictions become even more uncertain. The anticipated development of breeder reactors should favor an increasing percentage
of power generation from nuclear power sources. Regardless of the
growth of nuclear power generation, an increase in fossil fuel
power generation is predicted to the end of this century. Unfortunately the cleanest power source, hydroelectric, is severly
limited in its growth potential.

In the appraising the impact of central power plants upon air
pollution it is useful to further subdivide the fossil fuels according to type, Table 3. An increase in all categories, coal, oil,
and natural gas, is projected. No dramatic changes in the percentage use of each type of fossil fuel is noted with the exception
of the recent shift from the use of oil to natural gas. A slightly
increased role for coal is projected, dependent upon the rate of
introduction of nuclear power.

Table 1.   ANNUAL CONSUMPTION OF ENERGY IN THE UNITED STATES (1)

Sector	1947		1965		1980	
	10^{15}Btu	% total	10^{15}Btu	% total	10^{15}Btu	% total
Household and Commercial	6.8	20	11.4	21	17.2	19
Industrial	12.8	39	17.3	32	24.8	28
Transportation	8.8	26	12.8	24	21.7	25
Electric Utilities	4.4	13	11.0	20	24.4	28
Miscellaneous	.5	2	1.6	3	-	-
TOTAL	33.2	100	54.1	100	88.1	100

Known United States energy resources appear sufficient to sustain the fossil fuel consumption of the electric power industry
beyond the 1980 projection without alteration of distribution among
fuel sources, Table 4.

Table 2. ANNUAL ELECTRIC UTILITY GENERATION
IN THE UNITED STATES (1)

Power Source	1947 10^9kw-hr	1947 % total	1965 10^9kw-hr	1965 % total	1980 10^9kw-hr	1980 % total
Hydro	85	33	181	17	350	13
Nuclear	-	-	4	.4	458-723	16 - 26
Fossil Fuel	173	67	872	83	1941-1676	71 - 61
TOTAL	258	100	1057	100	2739	100

Table 3. ANNUAL USE OF FOSSIL FUELS IN ELECTRIC POWER
GENERATION IN THE UNITED STATES (1)

Fossil Fuel	1947 10^{15}Btu	1947 % total	1965 10^{15}Btu	1965 % total	1980 10^{15}Btu	1980 % Total
Coal	2.1	70	6.0	67	11.1 - 13.4	74-78
Oil	.5	17	.6	7	.9	6-5
Natural Gas	.4	13	2.4	26	3.0	20-17
TOTAL	3.0	100	9.0	100	15.0-17.3	100

Practically all present utilization of fossil fuels for electric power generation involves the production of steam and use of turbo-electric generators. Some use of gas turbine power is being developed primarily for peak load operation. Although greated use of gas turbines in power generation is predicted, especially as part of "total energy" concepts, they are not expected to represent a significant fraction of the total power generation capacity. Long term projections for the use of other generating methods, particularly direct energy conversion, to supplant the turbogenerator are uncertain. The most promising device is the "magnetohydrodynamic

(MHD) generator" which would provide for the direct utilization of
fossil fuel combustion products without steam production, turbines,
and rotating generators. Perhaps the most noteworthy development
in this field is the current construction of a 75 megawatt, natural
gas fired, combined magnetohydrodynamic-steam pilot power plant in
Moscow. The MHD generator accounts for 25 megawatts of the output.
The use of direct energy conversion as a major source of electric
power is not anticipated before the end of the century.

Table 4. UNITED STATES ENERGY RESOURCES (1962 PROJECTION) (1)

Resource	Known Reserve (presently recoverable) 10^{15}Btu	Total Projected Reserve 10^{15}Btu
Coal	4,600	88,600
Oil (liquid hydrocarbon)	300	3,900
Oil (shale)	-	945,000
Natural Gas	300	2,400-26,000
Uranium	300	24,000,000

    Projections indicate a continued increased utilization of fossil
fuels for power generation. Without reduction in combustion product
pollutants from central power plants, the contribution of these
plants to air pollution will continue to increase.

POWER PLANTS

    Most of the fossil fuel power generation in the United States
is in installations based on large furnace boilers which produce
hundreds of thousands of pounds of steam per hour per unit. These
furnaces are huge installations, a typical size being 20 to 30 feet
square by perhaps 80 to 100 feet in height. A typical coal fired
unit is shown in Figure 1. Design variations deal primarily with
the means and location of firing the coal. A typical oil or
natural gas fired boiler is shown in Figure 2. The same unit may
be used to burn fuel oil, natural gas, or fuel oil and natural
gas simultaneously.

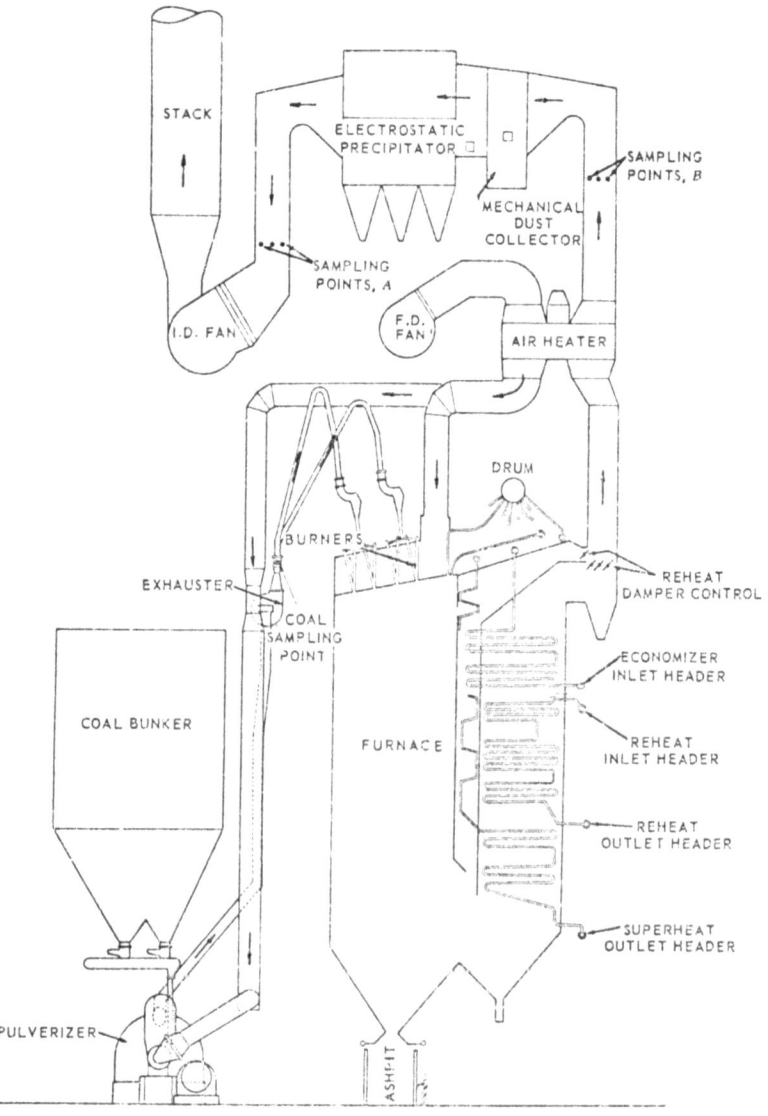

Figure 1. Coal fired steam generator, vertically fired (2).

Steam is heated in tubes which line the furnace wall. To im-
prove the efficiency of power generation, increased steam temperatures and pressures are required. Steam temperatures of 1000°F and
pressures of 2000 psia are common. Boiler tube corrosion is a
major problem. Air, usually preheated, is forced into the furnace
by large blowers. Pressure within the furnace is near atmospheric.

Power plants are operated with careful attention to maximizing efficiency. The low concentration of hydrocarbons and carbon monoxide found in power plant exhaust gases attest to the attainment of high combustion efficiencies. Unfortunately, as in other combustion processes, the converse is not true. The attainment of high combustion efficiency is not synonymous with low air pollutant emissions.

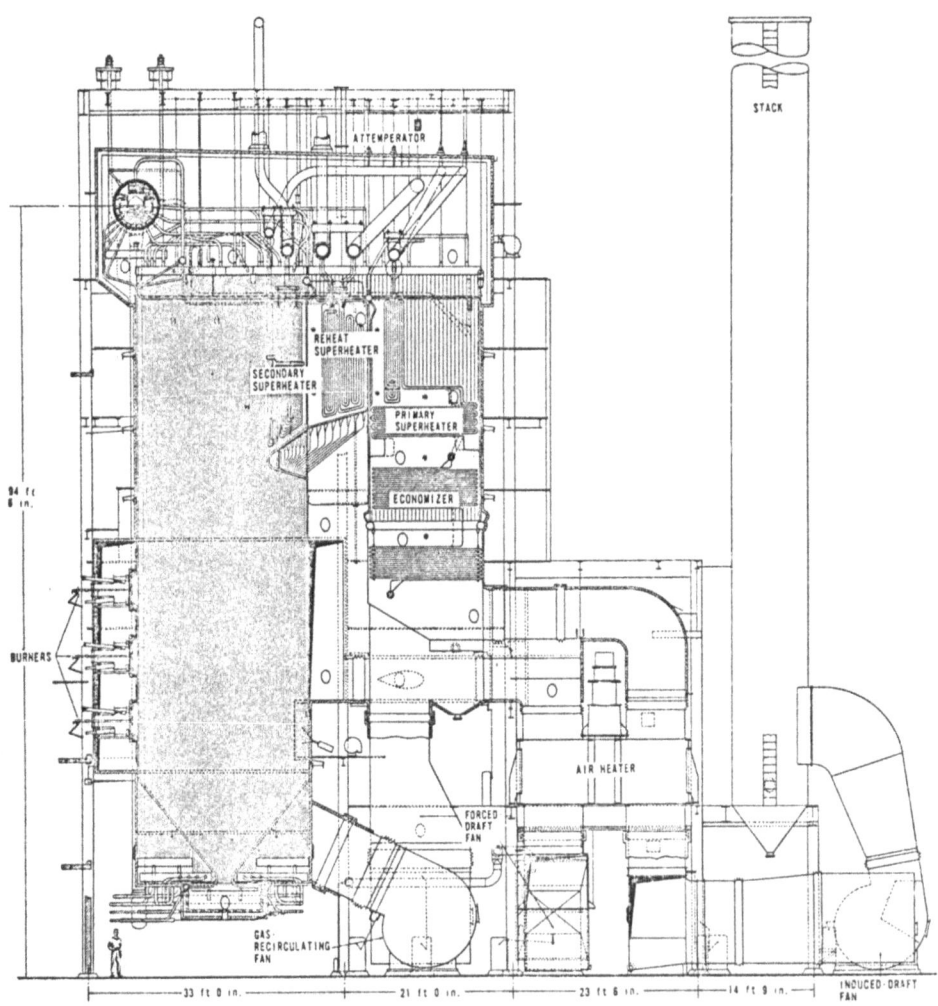

Figure 2. Oil or gas fired steam generator, front fired (3).

POLLUTION CHARACTERISTICS

The pollutants emitted from power plants depend largely upon the fuel burned and the furnace design. Average values of emissions based on fuel type are summarized in Table 5. For purposes of comparison, values are reported in terms of an "emission index" which relates the pollutants to the mass of fuel consumed.

Table 5.   Average air pollution emissions from power plants according to fuel type. Reported as an "emission index" defined as pound species per 1000 pounds of fuel. Calculated from reference 4.

Fuel	Particulate*	CO	HC	NO_x	SO_x	HCOH
Coal	85	.25	.1	10	38	.002
Oil	1.7	.07	.5	17	52	.1
Natural Gas	2.7	neg.	neg.	70	.1	.2

*without control

The contribution of power plants to air pollution is primarily particulates, oxides of nitrogen, and oxides of sulfur. Electric power plants in the United States account for about half of the sulfur dioxide emitted in the atmosphere and about one fourth of the particulates and the oxides of nitrogen. In local areas these figures may be significantly different. Carbon monoxide, hydrocarbon, and aldehyde emissions are low.

In the absence of emissions controls, the particulates and sulfur dioxide from power plants are related directly to the composition of the fuel burned. Oil and particularly coal contain ash which contributes to particulate emissions. Oil and coal contain sulfur which is emitted following combustion primarily as sulfur dioxide. The emission of nitric oxide depends primarily upon the maximum combustion temperature and is reported to be highest for natural gas fired furnaces. There is some dispute, however, over whether fuel oil or natural gas gives higher nitric oxide emissions. Since the maximum temperature is affected by the furnace and combustion processes as well as the fuel type, it is not unexpected that nitric oxide concentrations would vary considerably among different furnaces.

Some typical coal compositions are recorded in Table 6.

Table 6.   ANALYSIS OF TYPICAL UNITED STATES COALS (2)

Component	Pa.	Ohio	W. Va. District 8	Ky. Strip	W. Va. Deep Mine	Ill.	Pa.	Ill.
Proximate Analysis (as fired), %								
Moisture	1.1	2.8	1.8	2.3	1.2	4.1	1.1	2.0
Volatile matter	30.8	37.2	32.9	38.3	36.2	42.9	37.0	36.5
Fixed carbon	48.3	44.0	53.6	19.6	54.5	44.8	54.5	53.6
Ash	19.8	16.0	11.7	9.8	8.1	8.2	7.4	7.9
Ultimate analysis (as fired), %								
Hydrogen	4.6	5.0	4.8	5.1	5.1	5.5	5.2	5.1
Carbon	65.8	64.2	72.1	70.9	70.0	77.4	77.4	73.7
Nitrogen	1.4	1.3	1.4	1.5	1.5	1.4	1.4	1.6
Oxygen	6.1	11.8	9.0	10.4	8.2	12.4	6.1	9.4
Sulfur	2.3	1.8	1.0	2.3	1.2	2.5	2.4	2.3
Ash	19.8	16.0	11.7	9.8	8.1	8.2	7.5	7.9
Heating value, Btu/lb	11,820	11,480	12,645	12,640	13,540	12,650	13,910	13,195

The average sulfur content of coal used for electric power in the United States is about 2.5 per cent. Sulfur originating in the fuel appears primarily as sulfur dioxide in the flue gas with smaller quantities of sulfur trioxide also present. Exhaust gas compositions for different boilers, coal types, and operating conditions are reported in Table 7. A general correlation between sulfur dioxide and coal sulfur concentration is noted. The term "low sulfur coal" is used to refer to coal with less than one per cent sulfur. About one-third of the current United States coal production is low sulfur coal.

Residual fuel oils contain from about .7 to 5.5% sulfur. Less than one-fourth of the residual oil processed in the United States contains less than one per cent sulfur.

Natural gas is typically low in sulfur and ash content. From the standpoint of air pollution, natural gas is the cleanest of the fossil fuels.

CONTROL TECHNOLOGY

A number of approaches to the control of central power plant air pollution are possible and should be considered. Perhaps the most promising long term solution is the substitution of nuclear power. The cost of nuclear power is currently estimated to be comparable to or below that of coal fired units. Construction costs are estimated to be comparable while fuel costs are lower for nuclear installations. Nuclear fuel supplies are expected to be adequate. With the adoption of air pollution controls upon fossil fuel power plants, the construction of nuclear power plants may be accelerated.

The removal of power plants from population areas promises to relax but not eliminate air pollution emission requirements. Mine-mouth siting is an attractive approach to reducing the impact of coal transportation costs which may also ease the air pollution burden in the cities. The selection of sites must be made with the avialability of cooling water in mind. Offshore siting of power plants has been proposed but no plans now exist for building such installations.

The reduction of sulfur dioxide can be approached in two different ways: (1) use of natural low sulfur or processing of the fuel to remove sulfur and (2) processing of the flue gas to remove sulfur dioxide. Supplies of low sulfur fuels are not sufficient to meet the demands placed by air pollution considerations. Sulfur in coal appears in two forms, as iron sulfide or in organic compounds. A number of physical, chemical, and even biological processes for the reduction of sulfur in coal have been proposed. It is possible to remove only part of the pyrite (iron sulfide) from

Table 7.   TEST CONDITIONS AND MAJOR POLLUTANT EMISSIONS (2)

| | Coal | | Fly-ash gr/scf[e] | | Emissions ppm by volume, dry basis | | | | | |
| | | | | | Nitrogen oxides[f] | | Sulfur dioxide | | Sulfur Trioxide | |
Type of boiler firing	Sulfur,[c] %	Ash,[d] %	B[g]	A[h]	B[g]	A[h]	B[g]	A[h]	B[g]	A[h]
Full-load tests[a]										
Vertical	2.9	20.2	4.8	0.18	221	310	1450	1730	66	9
Corner	1.7	14.9	3.7	0.23	526	413	1150	1130	8	12
Front-wall	2.3	10.3	2.5	0.44	416	606	2120	1680	11	7
Spreader-stoker	2.8	8.4	2.3	0.38	431	437	1380	1570	58	76
Cyclone	2.4	7.7	1.5	0.39	1204	1160	1350	1360	21	31
Horizontally opposed	2.4	8.2	4.9	0.68	393	350	1560	1380	10	9
Partial-load tests[b]										
Vertical	2.8	19.0	4.7	0.11	161	171	1700	1640	46	10
Corner	1.6	13.5	2.9	0.13	393	325	1120	1000	10	12
Front-wall	1.8	9.2	2.4	0.22	500	453	1080	1460	3	20
Spreader-stoker	2.5	8.7	1.5	0.19	430	390	1280	1240	52	69
Cyclone	2.4	7.4	1.8	0.22	742	784	1380	1370	13	22
Horizontally opposed	2.9	7.8	2.9	0.61	395	328	1780	1680	6	8

[a]Average values from three or four tests.
[b]Average values from two tests.
[c]Moisture-and ash-free basis.
[d]Moisture-free basis.
[e]Corrected to 12% CO_2, dry basis.
[f]Reported as NO_2.
[g]Before fly-ash collector.
[h]After fly-ash collector.

coal. Economic considerations based on current technology favor
treatment of the flue gas over sulfur removal from coal. Low
sulfur residual fuel oil may be obtained by processing low sulfur
crude, desulfurization of crude, and blending of low and high
sulfur fuels. A number of desulfurization processes have been
proposed and plants for this purpose have been built and operated
successfully.

The substitution of exotic or unusual means of power generation is not likely to be of practical interest before 1980 and
perhaps before the end of the century. Concepts, in addition to
magnetohydrodynamic conversion, include utilization of solar radiation, geothermal energy, wind power, tidal power, thermionic
devices, thermoelectric devices, and fuel cells.

The use of high stacks disperses air pollutants over a wider
area thereby recuding the impact on a particular locality but increasing the exposed area. Stack heights in excess of 1000 feet
have been designed.

The removal of particulates and smoke from stack gases received early attention. As a result, techniques for control of
fly ash are both technologically and economically feasible. Three
general types of dust collecting are employed. In cyclone separators particulare matter is removed from the flue gas stream by
centrifugal force. In wet scrubbers the particulate matter is
removed by trapping them in a liquid spray. Electrostatic precipitators remove particulates by first depositing an electrostatic charge on the particle then collecting the particles on
plates of opposite polarity. The use of bag filters is generally
not successful because of problems of high pressure losses, high
temperatures, and corrosive flue gases. The efficiency of cyclone
collectors, about 70 per cent, may not be acceptable as more
strigent regulations are invoked. Electrostatic precipitators
with efficiencies in the 99 per cent range are possible and are
probably the most attractive devices. Some fly-ash collection
data are presented in Table 8.

A large number of processes have been proposed for the removal of sulfur from the flue gases. In power plants the sulfur
is discharged primarily as sulfur dioxide (90 to 100 per cent).
A summary of these processes appears in Table 9.

Of these processes, the limestone process, Figure 3, is
the closest to commercial application. The process is relatively
simple and unlike the others does not require a large physical
plant and capital investment. The other processes all result in
a marketable product. If the dry process is used, larger electrostatic precipitators may be necessary. The wet process carries a

disadvantage arising from the presence of water soluble magnesium
sulfate as a reaction product which may cause disposal problems.
The ash disposal may be compounded by the added requirement to
dispose of the nonmarketable products of the dolomite process.

After particulates and sulfur dioxide, nitric oxide is the
third major pollutant from central power plants. Nitric oxide is
formed in high temperature and excess air environments. Power
plant stack gas concentration of nitric oxide range from about
300 to as high as 1200 parts per million. Once formed it is difficult to dissociate nitric oxide back to its elements, molecular
nitrogen and oxygen. Nitric oxide is more difficult to remove
than sulfur dioxide. Modification of the combustion process has
been successful in reducing nitric oxide emission, **Figure** 4.
Burner designs have been altered in existing furnaces through re-
distribution of the mixture ratio. In place of the usual uniform
mixture ratio for all burners, some burners are operated with
excess fuel. The air deficiency is made up either by operating
other burners lean or by adding the air at a later point.

Although the effect has been explained in terms of modifying
the cooling process to promote nitric oxide dissociation, it
appears likely that the increased radiation heat transfer resulting from a region of fuel rich combustion results in a lowering
of the maximum temperature. The nitric oxide problem has been
greatest with the use of natural gas as the fuel. Further refine-
ments of the furnace design to reduce nitric oxide formation
appear feasible.

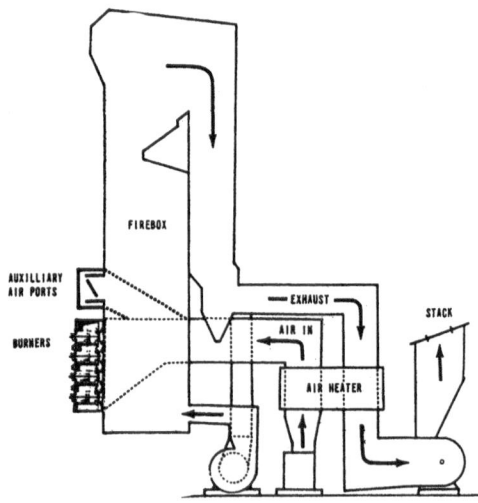

Figure 3.   The nonregenerative limestone process for scrubbing of
            sulfur oxides (2).

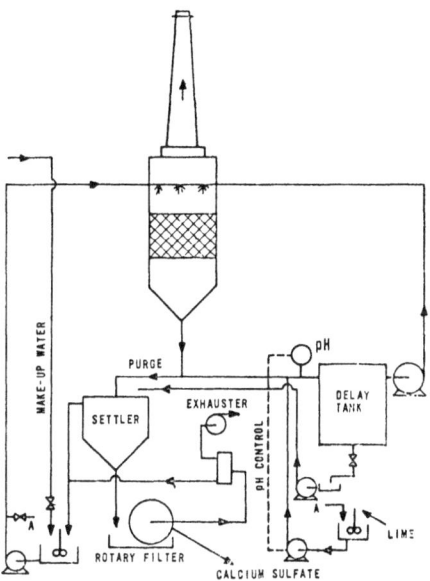

Figure 4. A front-fired boiler modified to provide two-stage
combustion (2).

The reduction of the flue gas oxygen content,i.e., operating
the furnace at a mixture ratio close to the stoichiometric value,
acts to reduce nitric oxide, Figure 5. Another proposal for
the reduction of nitric oxide is to recirculate some of the flue
gas through the furnace. This would act to reduce the maximum
temperature and thereby reduce the formation of nitric oxide.
This is comparable to exhaust gas recirculation in piston engines.
Difficulties in handling of the recirculated gases appear likely.
The reduction of furnace air preheat should have a similar effect,
at the expense of plant efficiency.

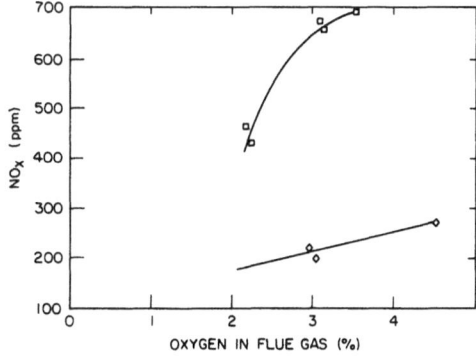

Figure 5. Effect of excess air on nitric oxide emission from oil
fired boilers (5).

Table 8. FLY-ASH CONCENTRATIONS AND COLLECTION EFFICIENCIES (2).

Type of boiler firing	Ash in coal,[a] %	Concentrations				Type of fly-ash collector[d]	Collector efficiency, %
		gr/scf[b]		lbs/1000 lb[c] dry flue gas			
		Bg	Ah	Bg	Ah		
Full-load tests[e]							
Vertical	20.2	4.8	0.18	8.8	0.27	C, E	96.4
Corner	14.9	3.7	0.23	6.9	0.42	C, E	93.9
Front-wall	10.3	2.5	0.44	4.6	0.82	E	83.1
Spreader-stoker	8.4	2.3	0.38	4.2	0.66	C	83.9
Cyclone	7.7	1.5	0.39	2.8	0.62	E	74.5
Horizontally opposed	8.2	4.9	0.68	8.9	1.27	C	83.9
Partial-load tests[f]							
Vertical	19.0	4.7	0.11	8.7	0.21	C, E	97.5
Corner	13.5	2.9	0.13	5.5	0.21	C, E	95.7
Front-wall	9.2	2.4	0.22	4.4	0.41	E	91.3
Spreader-stoker	8.7	1.5	0.19	2.8	0.35	C	87.3
Cyclone	7.4	1.8	0.22	3.1	0.36	E	86.3
Horizontally opposed	7.8	2.9	0.61	5.1	1.1	C	77.7

[a]Moisture-free basis.

[b]Corrected to 12 percent CO_2, dry volume basis.

[c]1000 pounds of dry flue gas corrected to 50 percent excess air.

[d]C designates cyclone: E designates electrostatic precipitator.

[e]Average values for either three or four tests of each unit.

[f]Average values for two tests at each unit.

[g]Before fly-ash collector.

[h]After fly-ash collector.

Table 9.   PROCESSES FOR THE REMOVAL OF SULFUR DIOXIDE FROM
FLUE GASES (Compiled from Reference 1).

Process	Technique	Est. Cost, $/ton Coal (with credit for by-product)
(1) Reinluft	Adsorption on activated charcoal	1.30
(2) Alkalized Alumina	Reactive Absorbent ($NaOH-Al_2O_3$)	.32-.86
(3) DAP-Mn	Reactive Absorbent (Mn_2O_3)	1.10
(4) Catalytic Oxidation	Vanadium Pentoxide Oxidation to Sulfur Trioxide, Removal by Condensation	.38-.72
(5) Kiyoura-TIT	Catalytic Oxidation-Ammonia Addition	.44
(6) Limestone	Reaction with Calcium Carbonate, dry or wet	.36-.63
(7) Wellman-Lord	Catalytic Oxidation (?) (secret process)	(-).60-0

SUMMARY

The continued growth of the electric power industry will be
accomplished with an increased use of fossil fuels. Without reductions in air pollutant concentrations in power plant flue
gases, power plants will become an increasing source of air
pollution. The primary pollutants are particulates, sulfur
dioxide, and nitric oxide. Techniques are available for the
removal of particulates. The removal of sulfur dioxide may be
accomplished either by use of a low sulfur fuel or by treating
the flue gas. The latter appears to be necessary but the
optimum method has not yet been identified. Reduction of nitric
oxide appears possible through modification of the furnace
combustion process. The design guidelines to accomplish a low
nitric oxide furnace are not well established.

References

1. Federal Power Commission Staff Report, "Air Pollution and
    the Regulated Electric Power and Natural Gas Industries,"
    Washington, D. C., September 1968.

2. Cuffe, S. T. and Gerstle, R. W., "Emissions from Coal-Fired
    Power Plants: a Comprehensive Summary," U. S. Dept. of
    Health, Education, and Welfare, PHS Publication No. 999-AP-
    35, 1967.

3. Danielson, R. A. (Editor), "Air Pollution Engineering Manual,"
    U. S. Dept. of Health, Education, and Welfare, PHS Publica-
    tion No. 999-AP-40, 1967.

4. Duprey, R. L., "Compilation of Air Pollutant Emission
    Factors," U. S. Department of Health, Education, and
    Welfare, PHS Publication No. 999-AP-42, 1968.

5. Bartok, W., et al., "Systems Study of Nitrogen Oxide Con-
    trol Methods for Stationary Sources," Esso Research and
    Engineering Company, Interim Status Report, Contract No.
    PH-22-68-55, May, 1, 1969.

COMBUSTION EMISSIONS FROM STATIONARY SOURCES

Milton Feldstein

Bay Area Air Pollution Control District

San Francisco, California

Emissions to the atmosphere from combustion of organic substances represent a substantial contribution of pollutants to the atmosphere. In this section we shall be concerned with emissions from combustion of fuel for space heating, water heating, industrial steam generation, and incineration of solid waste, agricultural burning and fireplace emissions. Combustion of fuel for power generation has been considered previously, and will not be included in the discussion.

Major pollutants emitted from these operations include particulate matter, nitrogen oxides, sulfur oxides, carbon monoxide and organic compounds. The air pollution effects of each of these are related to the quantity emitted by the operations under discussion. A brief review of the air pollution effects of each contaminant is in order at this point to set the stage for assessment of the contribution of each category to the overall air pollution problem in a given area.

Particulate Matter

The following table summarizes air pollution effects associated with the emission of particulate matter from combustion operations.

Particle Size		Effect
0.1 to 1.0 μ	(a)	Visible plumes
	(b)	Visibility reduction
	(c)	Lung deposition and health effects
	(d)	Suspension in atmosphere
> 10 μ	(a)	Dust fall
	(b)	Soiling

Carbon Monoxide
     Major concern is with health effects

Nitrogen Oxides
     Air pollution effects include:
          (a)   coloration of the atmosphere
          (b)   photochemical reactions
          (c)   vegetation damage
          (d)   health effects

Sulfur Oxides
          (a)   vegetation damage
          (b)   health effects
          (c)   visibility reduction after participation in the
photochemical process
          (d)   corrosion (SO_3)
          (e)   odor

Organic gases
          (a)   odor
          (b)   photochemical reaction
          (c)   eye irritation (aldehydes)

     These effects have been discussed in earlier lectures and are
listed here only to serve as a review. We shall next consider the
contribution of various classes of combustion categories to the
total atmospheric load of contaminants. (1)

I Gas Burning Sources

     Gas fired equipment generally produces a minimum of air pollu-
tants. Improper operation of air-fuel ratios can markedly increase
the emission of all 5 primary pollutants. Most marked increase
occurs when air is restricted, resulting in the emission of carbon
black or soot in dense black plumes.

     Table 1 summarizes emissions from natural gas burning in the
Bay Area from domestic use, Table 2 from commercial use and Table
3 from industrial use. The tables list the emission factors used
in the Bay Area, and the tonnage emissions are for the 6-county
Bay Area Air Pollution Control District. Total emissions for gas
combustion are summarized in Table 3 and show nitrogen oxides and
organic gases as the largest quantities of pollutants for this category.

II Fuel Oil Combustion

     Combustion of fuel oil produces significantly more quantities
of sulfur oxides and particulate matter than natural gas consumption. Fuel oils vary in sulfur content and ash, each having a re-

lationship to the quantity of pollutant produced. A summary of emissions from fuel oil burning boilers in the Bay Area is shown in Table 4. In order to put these data in perspective with power plant emissions, Table 5 is included to show emissions from power plants in the Bay Area.

Total emissions from the combustion of fuel oil and natural gas are shown in Table 6.

III Solid Fuel Combustion

While the combustion of solid fuels is almost non-existent in the Bay Area, it is an important source of air pollutants in many areas of the world. Major pollutants include particulate matter and sulfur oxides. Particulate matter consists mainly of carbon, silica, alumina and iron oxide. In addition to sulfur oxides, gaseous emissions include organic compounds, nitrogen oxides and carbon monoxide. Some emission data are shown in Table 7. Additional data on polynuclear hydrocarbon emissions for coal fired furnaces are shown in Table 8.

IV Incineration

The disposal of organic waste material generated in urban areas is a growing problem. The per capita quantity of waste continues to rise, making the problem more acute. Combustion of organic wastes gives rise to significant quantities of air pollutants, primarily organic compounds, particulates and carbon monoxide. Emission rates and factors are shown in Table 9 for various types of incinerator processes in the Bay Area. It would be of interest to compare emissions from municipal incinerators with figures shown in Table 9:

Municipal Incinerator Emissions[2]

Organics	1.2 pounds/ton
Carbon monoxide	1.0 pounds/ton
Sulfur oxides	2.0 pounds/ton
Nitrogen oxides	2.0 pounds/ton
Particulates	17.0 pounds/ton

V Agricultural Burning

Disposal of agricultural wastes presents many problems of concern to air pollution officials. Much the largest percentage of prunings, stubble and straw continues to be disposed of by burning in the open. There is some argument about the relative contribution of combustion contaminants from this source. Data shown in Table 10 summarize emissions in the Bay Area. (3) Included in the agricultural burning are burning for land clearing and variances for special meteorological burns.

TABLE 1

EMISSIONS FROM NATURAL GAS COMBUSTION DOMESTIC USE

		No. of Units	Emission Factor	Fuel Consumption Gas MMSCFD	Emissions T/D
COMBUSTION					
NATURAL GAS -- DOMESTIC					
Space Heaters	Organics	1,290,000	105 lb/MMSCF	270.5	14.2
	NOx		100 lb/MMSCF		13.5
	SOx		0.432 lb/MMSCF		.06
	CO		0.4 lb/MMSCF		.05
	Particulate		18.7 lb/MMSCF		2.5
Water Heaters	Organics	1,260,000	0.1 lb/MMBTU at 1069 BTU/CF=	91.4	4.8
	NOx				4.9
	SOx		108 lb/MMSCF		0.02
	CO				0.02
	Particulate				0.8
Ranges	Organics	1,170,000		43.8	2.3
	NOx		59.3 lb/MMSCF		1.3
	SOx				0.01
	CO				0.01
	Particulate				0.4
Refrigerators	Organics	3.4			0.2
	NOx		30.6 lb/MMSCF		0.06
	SOx				Negl.
	CO				Negl.
	Particulate				0.04

TABLE 1 (continued)

EMISSIONS FROM NATURAL GAS COMBUSTION DOMESTIC USE

	No. of Units	Emission Factor	Fuel Consumption Gas MMSCFD	Emissions T/D
COMBUSTION **NATURAL GAS -- DOMESTIC**				
Clothes Driers Organics			0.4	0.02
NOx		43.8 lb/MMSCF		0.01
SOx				Negl.
CO				Negl.
Particulate				Negl.
Total -- Domestic Organics			409.5	21.5
NOx				19.8
SOx				0.1
CO				0.1
Particulate				3.7

TABLE 2

EMISSIONS FROM NATURAL GAS COMBUSTION COMMERCIAL USE

NATURAL GAS -- COMMERCIAL		No. of Units	Emission Factor	Fuel Consumption MMSCFD	Emissions T/D
Space Heaters	Organics	245,000	105 lb/MMSCF	99.2	5.2
	NOx		100 lb/MMSCF		5.0
	SOx		0.432 lb/MMSCF		0.02
	CO		0.4 lb/MMSCF		0.02
	Particulate		18.7 lb/MMSCF		0.9
Water Heaters	Organics	123,000	108 lb/MMSCF	10.3	0.5
	NOx				0.6
	SOx				Negl.
	CO				Negl.
	Particulate				0.1
Ranges	Organics	25,5000	59.7 lb/MMSCF	12.4	0.6
	NOx				0.4
	SOx				Negl.
	CO				Negl.
	Particulate				0.1
Clothes Driers	Organics	3,000	118 lb/MMSCF	6.4	0.3
	NOx				0.4
	SOx				Negl.
	CO				Negl.
	Particulate				0.06

TABLE 2 (continued)

EMISSIONS FROM NATURAL GAS COMBUSTION COMMERCIAL USE

NATURAL GAS -- COMMERCIAL		No. of Units	Emission Factor	Fuel Consumption MMSCFD	Emissions T/D
Miscellaneous Gas Appliances	Organics	13,000	75.2 lb/MMSCF	13.4	0.7
	NOx				0.5
	SOx				Negl.
	CO				Negl.
	Particulate				0.1
Total -- Commercial					
	Organics			141.5	7.3
	NOx				6.9
	SOx				0.02
	CO				0.02
	Particulate				1.3

In disagreement with these figures, Darley has presented data on agricultural burning which show much lower quantities of pollutants emitted. These data are summarized here:

Particulate	17 pounds/ton
Organics	13.2 pounds/ton*
Carbon monoxide	60 pounds/ton
Nitrogen oxides	2 pounds/ton

*includes organic acids

Most of this data was developed with a specially constructed silo facility designed to simulate open burning. Measurements were made on small quantities of material which do not simulate field conditions of open burning.

VI Fireplace Emissions

While no jurisdiction has yet acted to control or eliminate fireplace burning in private homes, it would be of interest to compare emissions from this source with overall combustion emissions. Tests made by the Bay Area Air Pollution Control District show the following average emissions for a series of fireplaces:

Particulates	30 pounds/ton
Organics	45 pounds/ton
Carbon monoxide	120 pounds/ton
Nitrogen oxides	0.1 pounds/ton

It is interesting to compare the emissions from dried firewood burning in a fireplace to the emissions reported by Darley for open agricultural pruning burning. It is difficult to estimate the number of fireplaces in use in the Bay Area, but using figures on backyard burning, we can estimate 700,000 units in the Bay Area and an average of 20 pounds of wood burned in each fireplace, to arrive at total emissions from fireplace burning during cold weather:

Particulates	105 tons/day
Organics	157 tons/day
Carbon monoxide	420 tons/day
Nitrogen oxides	0.4 tons/day

Table 6 summarizes the emissions from all combustion sources considered, and in Table 11, those are placed in perspective with total emissions from all sources in the Bay Area. We can see that stationary source combustion emissions account for 25 percent of the organics, 35 percent of the particulates, 16 percent of the carbon monoxide, 36 percent of the nitrogen oxides and 43 percent of the sulfur oxides.

TABLE 3

EMISSIONS FROM NATURAL GAS COMBUSTION INDUSTRIAL USE

INDUSTRIAL	No. of Units	Emission Factor	Fuel Consumption	Emissions T/D
Boilers (under 500 hp)				
Firm Gas Rate	2000		119.3	
Organics		165 lb/MMSCF		9.8
NOx		170 lb/MMSCF		10.1
SOx		0.432 lb/MMSCF		0.02
CO		0.4 lb/MMSCF		0.02
Particulate		17.5 lb/MMSCF		1.0

TOTAL EMISSIONS
(Gas Combustion)

Organics..	38.6
NOx..	36.8
SOx..	0.14
CO...	0.14
Particulate..	6.0

TABLE 4

EMISSION FROM BOILERS

	No. of Units	Emission Factor	Fuel Consumption Gas MMSCFD	Fuel Consumption Oil T/D	Emissions T/D
INDUSTRIAL					
Boilers (under 500 hp)					
Fuel Oil Only Organics	330	127 lb/1000 bbls= 0.72 lb/T		540	0.2
NOx		3060 lb/1000 Bbls-= 18 lb/T			4.9
SOx		7980 lb/1000 Bbls= 46 lb/T			12.4
CO		6.3 lb/1000 Bbls= 0.037 lb/T			0.01
Particulate		4.81 lb/T + 2.5 wt.% of SO_2 as SO_3			1.3 / .3
Boilers (under 500 hp)					
Interruptible Organics	1,400	165 lb/MMSCF	89.5		7.4
NOx		170 lb/MMSCF			7.6
SOx		0.432 lb/MMSCF			0.02
CO		0.4 lb/MMSCF			0.02
Particulate		17.5 lb/MMSCF			0.8
Large Boilers (over 500 hp) Organics	76	165 lb/MMSCF 0.72 lb/T	26.8	569	2.2 / 0.2
NOx		170 lb/MMSCF 18 lb/T			2.3 / 5.1

TABLE 4 (continued)

EMISSION FROM BOILERS

	No. of Units	Emission Factor	Fuel Consumption Gas MMSCFD	Oil T/D	Emissions T/D
SOx		0.432 lb/MMSCF 46 lb/T			Negl. 13.1
CO		0.4 lb/MMSCF 0.037 lb/T			Negl. 0.01
Particulate		17.5 lb/MMSCF 4.81 lb/T+2.5 Wt% of SO_2 as SO_3			0.2 1.4 .3
Sub-Total -- Industrial			235.6	1109	
Organics					19.8
NOx					30.0
SOx					25.5
CO					0.06
Particulate					5.3

TABLE 5

EMISSIONS FROM UTILITIES

	No. of Units	Emission Factor	Fuel Consumption Gas MMSCFD	Oil T/D	Emissions T/D
UTILITIES					
Power Plant Boilers	10 Plants		338	1653	
Organics		4 lb/MMSCF			0.7
		0.954 lb/T			0.8
NOx		400 lb/MMSCF			67.6
		28.9 lb/T			23.9
SOx		0.432 lb/MMSCF			0.07
		102.3 lb/T			84.6
CO		0.4 lb/MMSCF			0.07
		0.037 lb/T			0.03
Particulate		15 lb/MMSCF			2.5
		5.22 lb/T+2.5			4.3
		wt.% of SO_2 as SO_3			2.1
Refinery Steam Boilers and Process Heaters (See Under Petroleum Refining)					
Organics					5.2
NOx					37.3
SOx					71.2
CO					0.5
Particulate					8.3

TABLE 5 (continued)

EMISSIONS FROM UTILITIES

	No. of Units	Emission Factor	Fuel Consumption		Emissions T/D
			Gas MMSCFD	Oil T/D	
Sub-Total -- Utilities, Refinery Boilers, Process Heaters					
Organics					6.7
NOx					129.
SOx					156.
CO					0.6
Particulate					17.2

TABLE 6

COMBUSTION EMISSIONS T/D 1967

	PARTICULATES	ORGANICS	NOx	SOx	CO	TOTAL
COMBUSTION						
Gas-Domestic	3.7	21.5	19.8	0.1	0.1	
Commercial	1.3	7.3	6.9	0.02	0.02	
Industrial	2.0	19.4	20.0	0.04	0.04	
Utilities	2.5	0.7	67.6	0.07	0.07	
Refineries	1.6	4.4	17.8	5.9	0.44	
Oil-Industrial	3.3	0.4	10.0	25.5	0.02	
Utilities	6.4	0.8	23.9	84.6	0.03	
Refineries	6.6	0.8	19.5	65.3	0.05	
	27.4	55.3	186	182	0.77	452
INCINERATION						
Residential	25.9	272	0.1	1.1	648	
Commer. Single C	2.6	29.0	0.3	0.3	27.1	
Multi C	0.6	0.2	0.5	0.2	0.01	
Wood burning	0.2	0.04	0.1	0.05	2.6	
Open-dump burning	0	0	0	0	0	
	29.3	301	1.0	1.6	678	1011
AGRICULTURE						
Orchard & vineyard	10.2	106	0.04	0.26	256	
Stubble & straw	-	-	-	-	-	
Range brush	6.2	65	0.03	0.16	156	

TABLE 6 (continued)

COMBUSTION EMISSIONS T/D 1967

	PARTICULATES	ORGANICS	NOx	SOx	CO	TOTAL
Weed control	0.13	1.4	negl	negl	3.3	
Variances	0.06	0.68	negl	negl	1.6	
Land clearing	1.2	12.7	negl	0.03	30.6	
	17.8	186	0.07	0.5	447	474
	74.5	542.3	187	184.6	1125	1937
TOTAL						

Control of Combustion Emissions

The control problems associated with the combustion of natural gas, fuel oil and solid fuel is primarily a matter of mechanical design. (5) (6) (7) (8). The combustion of natural gas in well-operated equipment generally produces a minimum of air pollutants. However, if burning conditions are poor, contaminant emissions dramatically increase. This is particularly important in air-fuel ratios, and type of mixing of air and fuel in the combustion process. Even under ideal conditions combustion of gas results in the formation of small quantities of organic compounds, aldehydes, carbon monoxide, sulfur oxides and other substances. The total contribution to the community atmosphere is a function of the quantity of fuel burned in the area. Reference is made to Tables 1, 2 and 3.

Combustion of fuel oil generally results in the emission of more contaminants, particularly in the area of sulfur oxides and particulate matter. These emissions are dependent upon the sulfur and ash content of the fuel. Again, control is primarily based on the design of burning equipment (5) (6) (7) (8). If particulate emissions are too high to meet standards of atmospheric quality, it may be necessary to reduce them with the use of electrostatic precipitators or other dust collection equipment. Similarly, sulfur oxide concentrations may have to be reduced (8) by increase in stack height, removal of sulfur oxide emissions within the stack, or removal of sulfur from the fuel oil. The assessment of the contribution of fuel oil burning operations to the Bay Area atmosphere can readily be made by reference to Tables 4 and 5.

Solid fuel combustion generally results in the emission of the largest quantity of air pollutants, and again control is primarily achieved in the design of combustion equipment. If additional reduction of particulate and sulfur oxide emission is required to meet regulatory rules, it may be necessary to install ancillary equipment (8).

Incineration of solid waste results in the emission of significant quantities of pollutants. Single chamber incinerators provide very poor combustion, (9) (10), while multiple-chamber incinerators with auxiliary gas burners provide efficient combustion of waste material, (9) (10), and minimum quantities of emitted pollutants. Most of the test data vailable, however, have been concerned with relatively small multiple-chamber incinerators. There are no test data available on the emissions from municipal type incinerators of this design. It is reasonable to assume that such large incinerators which may be able to handle two or three thousand tons of solid waste per day could meet stringent air pollution control regulations. It probably would be necessary to add auxiliary dust collection equipment for such incinerators, including scrubbers, multi-

TABLE 7

EMISSIONS FROM COAL FIRED UNITS

	Power Plant	Industrial Boiler	Domestic Boiler
Particulates	150-200	150-200	-
Sulfur Oxides			
2% S	76	76	76
1% S	38	38	38
Nitrogen Oxides	20	20	8
Carbon Monoxide	0.5	3	50
Organics	0.2	1	10

Emission factors can be converted to
a BTU basis by using conversion factor
26×10^6 BTU released/ton of coal

TABLE 8

POLYNUCLEAR HYDROCARBON
EMISSIONS FROM COAL FIELD
FURNACES (Pounds/MM BTU Input)

	Pulverized Firing	Chain Grate Stoker	Hand Fired
Benzo-a-pyrene	0.04-0.13	0.082	880
Pyrene	0.20-0.40	0.860	1320
Benzo-e-Pyrene	0-0.58	0.290	220
Perylene	0-0.15	-	132
Fluoranthene	-	1.50	2200

TABLE 9

EMISSIONS FROM INCINERATION OF WASTE

		No. of Units	Emission Factor	Refuse Burned T/D	Emissions T/D
INCINERATION					
Residential (Back Yard)	Organics	728,000	252 lb/T	2160	272
	NOx		0.1 lb/T		0.1
	SOx		1 lb/T		1.1
	CO		600 lb/T		648
	Particulate		24 lb/T		25.9
Commercial Single Chamber (Incl. Apartment houses)	Organics	1,300	90 lb/T	645	29.0
	NOx		1.0 lb/T		0.3
	SOx		1.0 lb/T		0.3
	CO		84 lb/T		27.1
	Particulate		8 lb/T		2.6
Multiple Chamber	Organics	590	0.8 lb/T	470	0.2
	NOx		2 lb/T		0.5
	SOx		1 lb/T		0.2
	CO		0.05 lb/T		0.01
	Particulate		2.51 lb/T		0.6
Wood Burners Silos	Organics	25	0.8 lb/T	100	0.04
	NOx		2 lb/T		0.01
	SOx		1 lb/T		0.05
	CO		53 lb/T		2.6
	Particulate		4 lb/T		0.2
Teepees		0	Eliminated during 1965		

TABLE 9 (continued)

EMISSIONS FROM INCINERATION OF WASTE

	No. of Units	Emission Factor	Refuse Burned T/D	Emissions T/D
Sub-Total -- Incineration				
Organics				301
NOx				1.0
SOx				1.6
CO				678
Particulate				29.3
				1011

TABLE 10

EMISSIONS FROM AGRICULTURAL BURNING

CATEGORY	Quantity Burned T/Yr	Emission Factor Lbs/T		Emissions T/D
Prunings Attrition	87991	Organics	250	106.3
		NOx	0.1	0.03
		SOx	0.6	0.26
		CO	600	256
		Particulate	24	10.2
Stubble, Straw Range, Weeds	62800	Organics	0.125	69.0
		NOx	0.00005	0.06
		SOx	0.0003	0.17
		CO	0.3	163.0
		Particulate	0.012	7.2
Variances	2000		Organics	.7
			NOx	-
			SOx	-
			CO	1.6
			Particulate	.06

TABLE 10 (continued)

EMISSIONS FROM AGRICULTURAL BURNING

CATEGORY	Quantity Burned T/Yr	Emission Factor Lbs/T		Emissions T/D
Land Clearing	37200		Organics	12.7
			NOx	-
			SOx	0.3
			CO	20.6
			Particulate	1.2
Total	189,991		Organics	186
			NOx	0.07
			SOx	0.5
			CO	447
			Particulate	17.8

TABLE 11

TOTAL EMISSIONS - BAY AREA (TONS/DAY)

Source	Particulates	Organics	Nitrogen Oxides	Sulfur Oxides	Carbon Monoxide
Combustion ⎱ Incineration ⎰ Agriculture ⎰	74	542	187	184	1125
Petroleum	2.4	187	9.4	57	443
Chemical	30	28	0.7	78	8
Metallurgical	26	2.5	7.6	75	0.6
Solvent Users	5.8	327	0.1	-	-
Ships and Railroads	6.7	6.1	7.7	9.5	2.2
Aircraft	8.8	8.3	6.9	-	35
Tractors and Construc- tion Equipment	19	31	16	5.1	68

TABLE 11 (continued)

TOTAL EMISSIONS – BAY AREA (TONS/DAY)

Source	Particulates	Organics	Nitrogen Oxides	Sulfur Oxides	Carbon Monoxide
Motor Vehicles	38	1027	279	20	5020
TOTAL	210	2159	514	429	6903

cyclones or electrostatic precipitators, all of which have been
employed successfully (11).

Some special purpose incinerators if not properly designed may
result in excessive emissions to the atmosphere. These include
sawdust burners, metal reclaiming incinerators, flue-fed apartment
house incinerators, and automobile body-burning incinerators. Solution to emission problems is related to design of equipment as a
primary method, and installation of dust collection equipment in
certain areas (6).

Regulatory Control of Combustion Emissions

Regulation 2 of the Bay Area Air Pollution Control District was
developed in 1960 to control the emission of particulate matter,
organic compounds and sulfur oxides from industrial and commerical
operations (12). It became effective in 1961 and enforcement of
this Regulation has been quite effective. Controls prescribed by
the Bay Area Air Pollution Control District are in the form of emission standards. The District is not permitted to specify the kind
of equipment that can be used to control emissions, nor is it permitted to issue permits for construction of operation of industrial
or commercial establishments. It can only set performance standards
on the emission of pollutants. These performance standards include
opacity requirements on smoke plumes, particulate and sulfur oxide
emissions and organic emissions from incinerators.

It would perhaps be of interest to discuss at this point what
is meant by performance standards or emission limits. In many air
pollution jurisdictions, control of emissions is attained by specifying particular types of equipment which must be installed to reduce the emission of solids or gases from industrial or commercial
sources. In the Bay Area the law (13) specifically prohibits the
District from specifying types of equipment used to achieve control.
Instead, emission limits are established and are applicable to all
operations which fall under that particular set of standards.

Emission of solid materials is controlled by two types of
standards. The first of these sets a limit on the opacity of smoke
being emitted from an industrial stack. One of the methods of limiting the emission of smoke and dust from industrial stacks is to
require that the opacity or density of the smoke be within a certain
prescribed range. A system which has been used for many years, and
referred to as the Ringelmann system, has been very useful in measuring emissions of smoke and dust from industrial stacks. Emissions
of black smoke from combustion operations were divided into categories based on the degree of blackness of the smoke. Thus, an
industrial stack emitting no smoke was assigned a Ringelmann 0, and
a stack emitting totally black smoke was assigned a Ringelmann 5.

Intermediate shades of black were assigned numbers 1,2,3, and 4.
A Ringelmann 1, for example, was a light grey color, and Ringelmann
2 a little bit darker. One could gauge the Ringelmann number by
reference to a series of charts equivalent to Ringelmann 1,2,3,4,
and 5 smoke.

Control or limitation of the quantity of smoke being emitted
is achieved by requiring that smoke of no darker shade than Ringelmann 2 could be emitted from a stack. If the stack emitted smoke
darker than Ringelmann 2, obviously more material was being emitted,
and the process was in violation of the Ringelmann requirement.

This system has been extended from the original concept for
black smoke to non-black smoke of equivalent obscuration as caused
by black smoke. In this manner any dust or smoke which obscures
the background to the same degree that black smoke does, would have
the same Ringelmann number as the black smoke.

The use of the Ringelmann chart for black smoke, and equivalent
opacity for non-black smoke has given control agencies a tool to
monitor such emissions from industrial plants. It is a convenient
tool in that an inspector can observe a plume at some distance from
the stack, and does not have to enter the premises in order to make
a reading. By limiting the Ringelmann number or the equivalent
opacity to No. 2 on the Ringelmann scale, the quantity of smoke and
dust being emitted from an industrial operation is thus controlled.
Obviously, a Ringelmann 3 plume will put out more smoke and dust than
a Ringelmann 2 plume.

Another method for controlling the quantity of solid material
emitted from industrial operations is to set a limit on the total
amount that can be emitted each day. Regulation 2 achieves this
by permitting no more than 0.3 grains of solid material per cubic
foot of gas to be emitted from certain industrial operations, and
no more than 0.2 grains per cubic foot from commercial incinerators.

Under Regulation 2, the most restrictive rule governing a
particular contaminant applies. Thus it is possible for an operation
to meet the 0.2 grain loading requirement and still be in violation
of the opacity requirement. The reason for this apparent anomaly
is due to the fact that the opacity of a plume is caused by the
light-scattering properties of very small particles of dust in that
plume. Because of their small size, they contribute very little
to the weight of material being emitted, yet they do contribute to
visibility reduction under certain meteorological conditions.

Other emission limits have been set for such contaminants as
sulfur dioxide, organic compounds and carbonyl compounds.

Sulfur dioxide emissions are controlled by restricting the amount emitted to 0.2 to v/v. However, a relatively new idea at the time, developed in the application of a ground level monitoring system for SO_2 emissions. Sulfur dioxide may cause problems when ground level concentrations approach certain values. These include damage to vegetation or health. A table of values of SO_2 concentrations and times at ground level was developed below which no effects were observed. The owner of a source of sulfur dioxide, which would include refineries and power plants, was required to monitor ground level concentrations and submit reports on the actual concentrations measured. If levels exceeded the limits which were set to safeguard health and vegetation, then a violation of the Regulation occurred. In this manner, the standard for SO_2 emissions (0.2%) could be exceeded so long as ground level concentrations were met. If ground level concentrations were exceeded, the emissions must be reduced to the original standard of 0.2% from the stack.

Regulatory control of incinerator emissions in the Bay Area include opacity requirements of Ringelmann 2 or less, grain loading requirements of 0.2 gr/5 DCF corrected to 6% oxygen, less than 50 ppm C_2 and higher hydrocarbons, and less than 50 ppm carbonyl compounds. Most of the multiple chamber incinerators tested easily meet these requirements (9) (10).

Particulate control for large process industries is achieved through the application of a process weight rule in Regulation 2. Allowable rates of emission increase with process weight. Because of the large source gas volumes associated with these industries, the grain loading requirement is generally much more stringent than

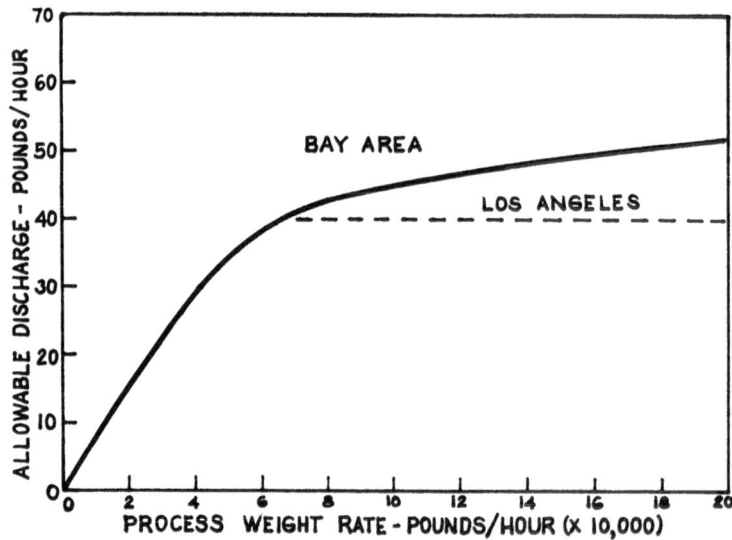

Figure 1. Process weight emission rates.

TABLE 12

REGULATORY CONTROL OF EMISSIONS FROM COMBUSTION OPERATIONS

Operation	Emission Limits				
	Particulate	Organics	SOx	NOx	CO
General Combustion					
Power, heat steam, etc.	0.3 gr/cf Process Wt. Opacity		0.2% or GIM*		
Incineration	0.2 gr/cf Opacity	50 ppm HC 50 ppm Ald.	0.2%		
Agricultural Burning	Meteorological control, seasonal, adequate mixing				
Backyard Burning	Banned, effective 1/1/1970				

*GIM = Ground level monitoring system for SO_2

the 0.3 grains/CF allowed for general operations. The process weight
emission rates are shown in Figure 1, and a summary of emission limits for particulate, organics, sulfur oxides and plume opacity is
shown in Table 12.

REFERENCES

1.   Source Inventory of Air Pollution Emissions in the Bay Area.
      Bay Area Air Pollution Control District, San Francisco, (1967).
2.   R. L. Duprey, Compilation of Air Pollutant Emission Factos.
      PHS Pub. No. 999-AP-42, (1968).
3.   Feldstein, M. et al, The Contribution of the Open Burning of
      Land Clearing Debris to Air Pollution. APCA Journal 13, 542,
      (1963).
4.   Darley, E., et al, Contribution of Burning of Agricultural
      Wastes to Photochemical Air Pollution. APCA Journal 16, 685,
      (1966).
5.   Engdahl, R. B., Stationary Combustion Sources in Air Pollution,
      A. C. Stern, ed, 2nd ed., Vol III, (1968).
6.   Air Pollution Engineering Manual, APCD, Los Angeles, PHS Pub.
      No. 999-AP-40 (1967).
7.   Control Techniques for Particulate Air Pollutants. NAPCA
      Pub. No. AP-51, (1969).
8.   Control Techniques for Sulfur Oxide Air Pollutants   NAPCA
      Pub. AP-52, (1969).
9.   Feldstein, M., Studies on the Analysis of Hydrocarbons from
      Incinerator Effluents with a Flame Ionization Detector.
      APCA Journal 12, 139, (1962).
10.   Tuttle, W. H., & Feldstein, M. Gas Chromatographic Analysis
      of Incinerator Effluents. APCA Journal 10, 427, (1960).
11.   Guide to the Appraisal and Control of Air Pollution. APHA,
      (1969).
12.   Regulations 1, 2 and 3, Bay Area Air Pollution Control District.
13.   Basic law creating Bay Area Pollution Control District.

COMBUSTION-GENERATED AIR POLLUTION

LEGAL ASPECTS

John A. Maga

California Air Resources Board

Sacramento

I. Introduction

The legal means to prevent and control air pollution have developed over a period of many years during which time much change has occurred in both the nature of the problem and legal basis for dealing with it. The first known anti-pollution laws were adopted in 1306 during the reign of King Edward II of England. These laws were directed against the use of sea coal because the burning of such coal caused smoke, soot and odor.

Shortly after 1900, air pollution was the subject of ordinances in a number of U. S. municipalities. Again, these laws dealt mainly with contaminants which were offensive to the human senses - smoke, soot, and odor with the burning of soft coal. This pattern of control of obvious pollutants through ordinances adopted by municipalities continued until quite recently.

A county-wide approach to air pollution control was taken by Los Angeles in 1947. This was one of the first attempts to include many communities in one control program rather than adopting local regulations, city by city. The latter half of the 1950's saw the establishment of laws and control programs at state and federal levels of government. In the past few years, many new government control agencies have been established, existing agencies were expanded, new laws enacted, and attention directed at all sources of air pollution.

II. Legal Basis for Preventing Air Pollution

A. Basic Rights of Individuals

Air pollution laws and regulations are society's effort to preserve the health and welfare by providing air of satisfactory quality. The need for such laws results from society's attempt to reconcile two basic and sometimes conflicting rights of individuals:

1. The right of a property owner to make use of his property.
2. The right of an individual and his property to not be damaged by the activities of others.

In rural areas and in an agricultural civilization, air pollution can be dealt with without extensive laws and numerous agencies. The industrial revolution and the growth of large metropolitan areas has made the problem much more complex. There are now many more pollutants and more people live in cities where they can be affected by these pollutants. The restrictions over the discharge of air contaminants has, therefore, also grown.

B. Common Law Nuisance

An individual can seek relief from air pollution in courts on the basis of the common law of nuisance.

A nuisance cannot be defined precisely. It usually is considered to be material annoyance, discomfort or harm produced by the unreasonable use of another person's property. This means that in each case it must be proven that the pollutants discharged were indeed injurious to property or offensive to the senses.

The common law makes a distinction between "private nuisance" and "public nuisance."

1. Private Nuisance

An individual who believes that his property is damaged can sue for an injunction or damages or both. He must show in each case that actual damage occurred, or at least, there there was substantial discomfort or inconvenience. It is clear that individuals will not normally seek relief under this law unless they suffered a considerable amount of damage or annoyance.

2. Public Nuisance

When a community or portion of a community is bothered by air pollution, a "public nuisance" may exist. An injunction suit can be brought by a public representative (city attorney, health officer, air pollution control officer) for the whole group.

Since court action in the absence of specific air pollution legislation provides a legal remedy for air pollution problems, it might be asked why there is a need for special air pollution laws? The answer is that legal remedy under the common law of nuisance has not answered the demands of the public for solutions to the problem of air pollution. Some of the reasons are:

1. The initiation and bringing to a successful conclusion of a nuisance action is often a lengthy and expensive undertaking. This almost always eliminates the private individual and discourages the public representative from beginning such action.

2. Even if the courts rule in favor of the affected people, the courts are not in a good position to know what remedy to require.

3. Nuisance action is not a practical way of preventing the discharge of contaminants from a large number of sources, to require solutions which are not obvious, or to correct problems caused by poor operation of facilities or equipment.

4. Nuisance action is not effective in preventing air pollution before the problems occur.

C. Police Powers

The state under its police power can enact laws specifically dealing with air pollution as a means of protecting the public health and welfare by insuring that the community has air of satisfactory quality. When a statute is enacted, it is no longer necessary to prove in court that each case is a nuisance. It is only necessary to show that the pollutant is discharged in excess of the quantity specified in the law or regulations. As a result, much of the doubt is removed on what constitutes air pollution. A basis also exists for making uniform decisions regarding the discharge of contaminants.

The states clearly have the authority to pass air pollution control laws. Local political bodies, such as a county or a city, may or may not have this authority, depending upon the state constitution or the laws which created the cities and counties. If the authority is not contained in the state constitution, the state can, of course, pass enabling acts which delegate the police power for controlling air pollution to the local government agencies.

III. Federal, State and Local Legislation and Programs

A. Federal

The first law providing for an air pollution program in the

federal government was enacted in 1955. The program was placed in the U. S. Public Health Service and included activities in the areas of research, training, and technical assistance to state and local agencies.

In the years since 1955, there have been several extensive amendments to the federal law. These have extended the authority of the federal government and greatly expanded the size and nature of its program. The federal activities now include research, training, air quality criteria, grants to encourage the establishment of new state and local programs and to support continuing programs, standards for motor vehicle emissions, air quality control regions, supervision of air quality standards programs for these regions, and abatement action in certain cases.

The federal government has become more and more concerned with the actual control of air pollution and much of its effort is now concerned with bringing about expanded control activities by the state and local agencies.

B. State

It has been only in the last 20 years that states have passed specific laws dealing with air pollution.

In 1947, California provided for the formation of county air pollution control districts. Prior to this, several states had passed laws permitting cities to control smoke, but there was no specific state law concerned with the entire problem of air pollution. Oregon in 1951, created the first state air pollution program. In 1955, California established a program at the state level of government. In the past few years, many more states have legislated air pollution control programs. The recent federal laws requiring the designation of air quality control regions by the Secretary of Health, Education and Welfare, and the adoption of air quality standards in the regions by the states will involve all states in air pollution control.

At this time, (1969), the states have a variety of air pollution laws which provide for some type of state control program. These programs vary widely in their organization, activities, and funding. Most state air pollution programs are in the health department. Some, however, have independent boards or commissions.

The role of the state usually falls between that of the federal and local governments. Local agencies are mainly concerned with enforcement; the federal government is mainly concerned with research, grants, and motor vehicle emission standards and the encouragement of state and local agencies. Normally, the federal government does not have extensive enforcement responsibilities.

States have major interests in coordination, investigations, air monitoring, technical assistance, and the establishment of standards. Although sometimes engaged in research, they have a minor role here as compared to the federal government. Most states rely on the local agencies for enforcement of laws and regulations, but retain some responsibilities in this area. Some of the states have major enforcement responsibilities.

C. Local

With the exception of motor vehicles, the enforcement of regulations against the discharge of contaminants is usually done by local agencies. Although these agencies are recognized as the ones which should do most of the enforcement work, it is not always clear what type of local agency - city, county, or regional authority - is best suited to carry out these responsibilities.

1. City Programs

The first air pollution control activities were to be found in cities. In 1910, about 20 of the larger U. S. communities held programs against smoke pollution. Even as late as 1960, a large majority, about 75 percent, of the control agencies were in city governments. However, the number of county, city-county, or regional programs had been increasing while there was little change in the number of city programs.

2. County Programs

Single community programs are limited in what they can do to insure satisfactory air quality when a large number of communities are located near enough to each other to be affected by common sources of air pollution. In these instances, it is difficult to have uniform regulations and enforcement in each of the cities and in the unincorporated area between the cities.

California attempted to deal with this problem by enacting in 1947 a state law which permitted the formation of county-wide air pollution control districts. Thus, one agency is responsible for air pollution control in all the cities as well as the unincorporated areas of the county. The Los Angeles County Air Pollution Control District, for example, includes approximately 80 incorporated cities.

County programs have been authorized and are in other states, such as Kentucky, Texas, and Florida.

3. Regional Programs

        As in the case of cities, a county often finds that its air
pollution problem is not confined to the county alone. Recently,
there has been interest in organizing multi-county or regional
control agencies.

        The first such agency was the San Francisco Bay Area Air Pol-
lution Control District which was created by the California State
Legislature in 1955. This district includes 6 counties and approximately 70 cities.

        In North Carolina, four counties have joined to establish one
program for the entire area. The Mid-Willamette Valley Air Pollution Control Authority includes five counties in Oregon. Several
other regional agencies have been formed or are being proposed in
other parts of the U. S.

        The Federal Clean Air Act of 1967 and the California Mulford-
Carrell Air Resources Act of 1967 further encouraged the control of
air pollution on a regional basis. The federal law provides for
the establishment of air quality control regions in the larger metropolitan areas of the U. S. Many of these have already been designated. The Mulford-Carrell Air Resources Act requires that California be divided into air basins. This was done by the Air Resources Board in November, 1968. Implementation of these laws
requires a coordinated approach to air pollution control in the
federal regions and in the air basins in California.

4. Interstate Programs

        Some metropolitan areas include portions of more than one
state. In these instances, it is difficult to have uniform laws
and enforcement because the authority for air pollution control is
located in two or more states and a number of local agencies.

        It has been suggested that compacts be used to achieve the
required uniform action. An interstate agency would be established
to administer the control program. This requires agreement by the
states if the agency is to be granted this authority. This agreement, in the form of a compact, must receive the consent of Congress.
The compact would then have the support of federal law and would
provide for enforcement by the included states.

        Several of the federal air quality control regions include
more than one state. This lends additional support to a coordinated control program in these areas.

IV. Air Pollution Control Programs

A. Nature of Legislation

The type of enabling legislation required to control emissions depends upon the organization desired and the powers to be given. The control program may include a number of activities - atmospheric monitoring, inventory of sources, adoption of rules and regulations, enforcement, and public information.

Enabling laws, therefore, must provide for all these activities and should:

1. Specify the conditions and procedures to be followed in activating the agency.
2. Designate the areas to be included.
3. Provide for the powers of the agency, including the authority to adopt and enforce rules and regulations and to carry out the provisions of the act.
4. Provide for the governing board, appointment of officers, and hiring of employees.
5. Include standards for the emissions of some of the more common pollutants.
6. Provide for funds required by the agency.
7. Provide for appeals from the administrative actions.
8. Establish the means of enforcing the emission standards, rules and regulations.

B. General Considerations

Restrictions against the discharge of specific pollutants may be included in the legislation or in the rules and regulations of the administrative agency. In general, the number of pollutants specifically restricted in the basic law should not be large.

The use of rules and regulations promulgated by the governing body of the control agency is the best way to deal with most sources of air pollution. This is particularly true when the air pollution problem is a complex one because of the presence of many sources and when there are not well established methods for controlling pollutants. The use of rules and regulations permits more flexibility to the control program than is the case with rigid requirements contained within the law itself.

Because the rules and regulations have force of law, provisions are made that in the process of their adoption, all individuals affected can be heard by the body who adopts the regulations. Public notice is required of the hearing at which the regulations are to be considered and adopted. Individuals opposing the regulations can give testimony and submit evidence to show the regulations are unreasonable or not necessary.

C. Motor Vehicle Emission Control

        Control of motor vehicle emissions has followed a different
pattern from that of the control of other sources. The Secretary
of Health, Education and Welfare (HEW), establishes emission standards for new cars which apply throughout the United States. The
individual states are prevented from enforcing motor vehicle emission standards for new cars but can do so for used cars. The
federal law, however, provides that a waiver can be granted to
California to set and enforce its own standards. To receive such
a waiver, however, California must show that it has an extraordinary
and compelling problem, its emission standards are more strict than
federal ones, and the emission standards are technologically feasible.

        The California Legislature has enacted extensive laws in this
area. The Pure Air Act of 1968 includes motor emission standards
which become progressively severe through the 1974 models. The
Secretary of HEW has granted California a waiver for these emission
standards.

V. Prospects for Future Legislation

        The public, mass media, local agencies, state legislatures,
and the U. S. Congress are showing a great interest in air pollution control. This problem has been the subject of much recent
legislation and can be expected to receive even more legislative
attention in the future. The future laws will be directed at
providing more strict control of air pollution, encouraging the
formation of new and more effective control agencies, and finding
methods to control pollutants that cannot be easily controlled at
this time.

A. Federal

        Federal programs will continue to stress research and support
of state and local programs. Emphasis will, in the future, probably
be placed on bringing about control of air pollution by:

1. Federal action in cases of interstate air pollution problems.
2. Grants to control agencies.
3. Encouraging or actually forcing state and local agencies to
undertake strong air pollution control programs. This could be
done by:

a. Requiring states to implement air quality and emission standards under the threat that if they do not, the federal government
will act.
b. Require the states to establish enforcement programs to insure

continued operation of vehicle emissions control systems.

B. State

Legislation at the state level will continue to vary a great
deal from state to state.

1. States will establish broader programs at the state level of
government because of the concern over the problem and because of
the need to establish and implement air quality standards in the
federal air quality control regions.
2. The states will become concerned with the control of air pollution throughout the entire state, rather than rely on local government to decide where and when to form control agencies as is usually
the case.
3. States will encourage the formation of regional control agencies.
4. Air quality and emission standards will become elements in the
state programs throughout the state rather than only in the federal
regions.

C. Local

Local air pollution activities and regulations can be expected
to result in:

1. Increasing numbers of control agencies.
2. Larger and more qualified staffs.
3. Organization of control programs on the basis of counties or
regions.
4. Adoption of regulations that are more restrictive and which
include more sources of air pollution.

References

1. California Health and Safety Code, Section 24198-24341 (County
      Air Pollution Districts); Section 24345-24374 (Bay Area Air
      Pollution Control District), Sections 3900-39570 (Mulford-
      Carrell Air Resources Act and Pure Air Act of 1968), State
      of California, Sacramento, California.
2. Clean Air Act of 1967, Public Law 90-148, 90th Congress.
3. Los Angeles County Air Pollution Control District, "APCD Rules
      and Regulations," Los Angeles, California.
4. S. Edelman. "Air Pollution Control Legislation," Chapter 50,
      AIR POLLUTION, Volume III, edited by A. C. Stern, Academic
      Press (1968).
5. J. J. Schueneman, "Air Pollution Control Administration,"
      Chapter 52, AIR POLLUTION, Volume III, edited by A. C. Stern,
      Academic Press (1968).
6. American Public Health Association, "Health Officials Guide to

Air Pollution," New York (1968).

7. Public Health Service, "A Digest of State Air Pollution Laws,"
    Public Health Service Publication No. 711, U. S. Department
    of Health, Education and Welfare, Washington, D. C.

8. D. G. Warren, "State and Local Regulations of Air Pollution,"
    Popular Government, February, 1967.

INDEX

329